N

Ernst Schering Research Foundation Workshop 33
Stem Cells from Cord Blood,
In Utero Stem Cell Development,
and Transplantation-Inclusive Gene Therapy

Springer
Berlin
Heidelberg
New York
Barcelona
Hong Kong
London
Milan
Paris
Singapore
Tokyo

Ernst Schering Research Foundation
Workshop 33

Stem Cells from Cord Blood, In Utero Stem Cell Development, and Transplantation-Inclusive Gene Therapy

W. Holzgreve, M. Lessl
Editors

With 27 Figures and 16 Tables

 Springer

Series Editors: G. Stock and M. Lessl

ISSN 0947-6075
ISBN 3-540-67701-1 Springer-Verlag Berlin Heidelberg New York

CIP data applied for

Die Deutsche Bibliothek – CIP-Einheitsaufnahme
Stem cells from cord blood, in utero stem cell development, and transplantation-inclusive gene therapy ; with tables / Ernst Schering Research Foundation. W. Holzgreve and M. Lessl, ed. - Berlin; Heidelberg; New York; Barcelona; Budapest; Hong Kong; London; Milan; Paris; Singapore; Tokyo: Springer, 2001
(Ernst Schering Research Foundation Workshop; 33)
ISBN 3-540-67701-1

Springer-Verlag Berlin Heidelberg New York
a member of BertelsmannSpringer Science+Business Media GmbH
© Springer-Verlag Berlin Heidelberg 2001
Printed in Germany

Typesetting: Data conversion by Springer-Verlag
Printing and Binding: Druckhaus Beltz, Hemsbach
SPIN:10773207   21/3130/AG–5 4 3 2 1 0 – Printed on acid-free paper

Preface

The title "Stem Cells from Cord Blood, In Utero Stem Cell Development, and Transplantation-Inclusive Gene Therapy" suggests that more than one topic is combined in one workshop. Indeed, at first glance the recovery of stem cells from cord blood has to be seen as separate from the attempts to achieve effective in utero therapy by stem cell transplantation, because the first issue deals with an innovative stem cell source as an alternative to bone marrow, which is already spreading rapidly in medical practice, whereas the second topic is still strictly experimental and only investigated in medical centers with the appropriate background. It is, however, not only justified, but helpful to combine the two topics in one workshop and consequently to cover them in the same volume of the Ernst Schering Research Foundation Workshop series, because they are intimately related and both based on the new insights into the biology of stem cells. Professor Werner Arber, the Nobel Laureate from the University of Basel, pointed out in his In-

Professor Dr. W. Holzgreve

The participants of the workshop

troductory Lecture that our understanding of hematopoietic stem cells
as descendents of totipotent cells and our current approaches to using
them in post- and prenatal therapy have been furthered significantly by
genetic engineering technologies which are "artificial contributions to
the process of biologic evolution". At the same time as the delineation
of the DNA base sequence of the human genome has been achieved,
more than 400 protocols for gene therapy have already been approved,
and recently proposals for an in utero approach were discussed at
length by the Recombinant DNA Advisory Committee (RAC) in the
USA, mainly stimulated by Dr. Anderson in California. We think that
in utero stem cell therapy is the logical step from the amazing progress
in prenatal diagnosis towards therapy as an alternative to abortion,
which parents have opted for so far after early in utero diagnosis of to-
date untreatable genetic diseases such as certain inborn errors of meta-

bolism. Sir William Liley, who developed techniques for in utero diagnosis and treatment of rhesus incompatibility and who therefore can be considered to be the "father of prenatal therapy", has rightfully urged researchers and physicians in the area of prenatal medicine to progress from a "search and destroy" to a "find and treat" mission. It is most likely that gene transfer to still-expanding stem cell populations will prove more successful than postnatal approaches with more committed cells. Since some viral vectors contain all the machinery for integration and can infect even non-dividing cells, it became obvious during the conference that stem cells in early fetuses can perhaps be transfected more easily and more successfully than stem cells post partum and that the new third generation lentiviral vectors hold great promise for in utero gene therapy. Biosafety issues, however, have to be taken into account, and tests in appropriate animal models need to be carried out to ensure that the manipulated cells do not reach the germ line, but are targeted towards the bone marrow in the case of hematopoietic stem cells. Since the paper by Bjornson et al. in *Science* in 1999 with the title "Turning brain into blood", the potential of stem cells has become known not only to scientists in the field, but also to the public, which demands from us clear statements on the benefits and risks of ultimately using these cells in medicine.

The use of stem cells from cord blood as an alternative to bone marrow, first reported by Broxmayr et al. (1989), is a success story of modern day medicine, because even the first child so treated is still alive more than 10 years later, and the New York Blood Foundation cord blood bank alone now has more than 10,000 units stored and around 1000 patients transplanted. The advantages of stem cells from cord blood are now clear in that the harvesting is easy and without medical problems for the donor, HLA mismatches are less important than in adult bone marrow, graft-versus-host disease is less likely and less severe, and international networks of cord blood banks would help to serve the populations which need transplants and which are underrepresented in bone marrow registries. The international organization Netcord now has registered already more than 50 banks worldwide with thousands of units frozen. During the workshop, the Professor of Law Kurt Seelmann from the University of Basel discussed the ethical and legal issues regarding the right of ownership of cord blood, consent, timing of cord clamping, and the link between bank and donor, which becomes important especially when investigation of the blood reveals unexpected findings.

In order to make stem cell transplantations, which are already well established for pediatric recipients, more suitable for adults, stem cell expansion techniques arc being sought, and work presented by Dr. Aleksandra Wodnar-Filipowicz and others at the conference has already shown achievements in expanding colony-forming units (progenitors) for over 10 weeks. Alternative sources for stem cells such as fetal liver were also discussed, but parallel to the biological advantages, ethical issues regarding these sources need to be discussed openly.

Prenatal therapy with stem cells has already been successful in human fetuses with severe combined immunodeficiency, who after early prenatal diagnosis and successful in utero application of stem cells were born with stable chimerism and are healthy, without any manifestations of the disease, so that they could even have their normal vaccinations. Microchimerism, which now seems to be a common feature of tissues early in prenatal life, can also be present postnatally with rare manifestations, for example, in autoimmune disorders. Furthermore, we have learned from experiments in monkeys that a partial state of tolerance, for example, to renal grafts, can be induced by in utero stem cell transplantation and that the stromal compartment seems to dictate the gene expression programs. Despite this amazing progress at the bench and already at the bedside in this field, many problems still need to be solved before in utero hematopoietic stem cell transplantation can become more widely available for effective therapy after prenatal diagnosis of, for example, metabolic disease or thalassemias.

We thank the Ernst Schering Research Foundation and especially Professors Raff and Stock for allowing us to dedicate the 33rd workshop in the series to this exciting topic, and Dr. Monika Lessl and Ursula Wanke for their excellent organization. The location of the workshop in the heart of Berlin (Berlin-Mitte), where the wall came down 10 years ago, could be interpreted as symbolic for the fact that seemingly unsolvable problems can be overcome with dedication and persistence. Regarding the use of "stem cells from cord blood, in utero stem cell development, and transplantation-inclusive gene therapy", despite the significant progress already achieved, equally significant problems still remain to be solved. The motto "proceed with caution" should apply.

Professor Dr. med. Dr. h.c. Wolfgang Holzgreve
Chairman, University Women's Hospital
Basel, Switzerland

References

Bjornson CRR, Rietze RL, Reynolds BA, Magli MC, Vescovi AL (1999) Turning brain into blood: a hematopoietic fate adopted by adult neural stem cells in vivo. Science 283:534–537

Broxmeyer H, Douglas G, Hangoc G, Cooper S, Bard J, English D, Arny M, Thomas L, Boyse E (1989) Human umbilical cord blood as a potential source of transplantable hematopoietic stem/progenitor cells. Proc Natl Acad Sci USA 86:3828

Introductory Remarks

The targeted use of stem cells and in particular of those committed to hematopoiesis is based on recent progress in both cell biology and molecular genetics. Gene technology, has opened novel approaches to the investigation of specific biological functions. It is the knowledge of the functional characteristics which permits a responsible development of new therapies such as the transplantation of specifically committed cells.

Genetic engineering includes steps of reflected genetic variation, and these can be seen as artificial contributions to the process of biological evolution. The risks of such approaches *should be carefully evaluated*, and a profound knowledge of the molecular nature of spontaneous genetic variation can be very helpful in this. Work with microorganisms has revealed that genetic variation results both from the activities of the products of specific evolution genes and from the influence exerted by non-genetic factors. Interestingly, some of the principles driving biological evolution are also known to be involved or are thought to be involved in cellular differentiation or commitment. These mechanisms include the reshuffling of DNA segments to produce functional genes, as in the generation of antibody-producing genes. Another presumptive role in cellular commitment is assigned to DNA methylation, which modifies the activities of DNA without changing its genetic information content. Molecular genetics can provide useful tools and can further explore the various strategies contributing to the specialization of the descendants of totipotent cells.

Prof. Werner Arber
Biocenter, University of Basel, Switzerland, Nobel Laureate

Table of Contents

List of Editors and Contributors

Editors

W. Holzgreve
Department of Obstetrics and Gynecology, University Hospital Basel,
Schanzenstrasse 46, 4031 Basel, Switzerland

M. Lessl
Ernst Schering Research Foundation, Müllerstrasse 178, 13342 Berlin,
Germany

Contributors

E.M. Anderson
Department of Immunology, Queen's Medical Centre,
Nottingham, NG7 2UH, UK

W. Arber
Abteilung Mikrobiologie, Biozentrum, Universität Basel,
Klingelbergstrasse 20, 4056 Basel, Switzerland

E. Buetti
Institute of Microbiology, University Hospital Lausanne, Switzerland

S.-H. Chou
Pediatric Bone Marrow Transplant Division, University of California,
San Francisco, CA 94143-1278, USA

T. Cook
Department of Histopathology, Imperial College School of Medicine, London, UK

E. Costello
Institute of Microbiology, University Hospital Lausanne, Switzerland

C. Coutelle
Cystic Fibrosis Gene Therapy Research Group,
Section of Molecular Genetics, Division of Biomedical Sciences,
Imperial College School of Medicine, London SW7 2AZ, UK

M.J. Cowan
Pediatric Bone Marrow Transplant Division, University of California,
San Francisco, CA 94143-1278, USA

A.-M. Douar
Genthon, Evry, France

G. Eichholz
Institut für Medizinische Strahlenkunde und Zellforschung,
University of Würzburg, Versbacherstrasse 5, 97078 Würzburg, Germany

J.H.F. Falkenburg
Department of Hematology, Leiden University Medical Center, K6-31,
2300 RC Leiden, The Netherlands

H. Geiger
Institut für Medizinische Strahlenkunde und Zellforschung,
University of Würzburg, Versbacherstrasse 5, 97078 Würzburg, Germany

M. Hanson
Cystic Fibrosis Gene Therapy Research Group,
Section of Molecular Genetics, Division of Biomedical Sciences,
Imperial College School of Medicine, London SW7 2AZ, UK

F. Harder
Institut für Medizinische Strahlenkunde und Zellforschung,
University of Würzburg, Versbacherstrasse 5, 97078 Würzburg, Germany

D.R.E. Jones
Department of Immunology, Queen's Medical Centre,
Nottingham, NG7 2UH, UK

H.H.H. Kanhai
Department of Obstetrics, Leiden University Medical Center, KG-31,
PO Box 9600, 2300 RC Leiden, The Netherlands

U. Kapp
University Medical Center, Hematology/Oncology,
Universitätsklinikum Freiburg, Hugstetterstrasse 55, 79106 Freiburg, Germany

T. Kiserud
Department of Obstetrics and Gynaecology,
Royal Free and University College Medical School, London UK

D.T.Y. Liu
Department of Obstetrics, Queen's Medical Centre, Nottingham, NG7 2UH,
UK

A. Luther-Wyrsch
Department of Research, University Hospital Basel, Basel, Switzerland

R. Mertelsmann
University Medical Center, Hematology/Oncology,
Universitätsklinikum Freiburg, Hugstetterstrasse 55, 79106 Freiburg, Germany

A.M. Müller
Institut für Medizinische Strahlenkunde und Zellforschung,
University of Würzburg, Versbacherstrasse 5, 97078 Würzburg, Germany

C. Nissen
Department of Research, University Hospital Basel, Basel, Switzerland

P. Orchard
Umbilical Cord Blood Transplantation Program, Pediatric BMT Program,
University of Minnesota, Saint Paul, MN 55108, USA

A. Pavirani
Transgene, Strasbourg, France

C. Rodeck
Cystic Fibrosis Gene Therapy Research Group,
Section of Molecular Genetics, Division of Biomedical Sciences,
Imperial College School of Medicine, London SW7 2AZ, UK

P. Rubinstein
Placental Blood Program, Fred H. Allen Laboratory, New York Blood Center,
310 East 67th Street, New York, NY 1002, USA

S. Scherjon
Department of Obstetrics, Leiden University Medical Center, KG-31,
PO Box 9600, 2300 RC Leiden, The Netherlands

H. Schneider
Cystic Fibrosis Gene Therapy Research Group,
Section of Molecular Genetics, Division of Biomedical Sciences,
Imperial College School of Medicine, London SW7 2AZ, UK

K. Seelmann
Institut für Rechtswissenschaften, University of Basel, Maigasse 51,
4052 Basel, Switzerland

L.E. Shields
Department of Obstetrics and Gynecology, Division of Perinatal Medicine,
University of Washington, Seattle, USA

C.E. Stevens
Placental Blood Program, Fred H. Allen Laboratory, New York Blood Center,
310 East 67th Street, New York, NY 1002, USA

D.V. Surbek
University of Obstetrics and Gynecology, University Hospital Basel,
Schanzenstrasse 46, 4031 Basel, Switzerland

A.F. Tarantal
California Regional Primate Research Center, Davis, California, USA

M. Thali
Institute of Microbiology, University Hospital Lausanne, Switzerland

M. Themis
Cystic Fibrosis Gene Therapy Research Group,
Section of Molecular Genetics, Division of Biomedical Sciences,
Imperial College School of Medicine, London SW7 2AZ, UK

D. Trono
Department of Genetics and Microbiology, Faculty of Medicine,
1 rue Michel-Servet, 1211 Geneva 4, Switzerland

R.M. Walker
Department of Animal Physiology, Queen's Medical Centre,
Nottingham, NG7 2UH, UK

M. Westergren
Department of Obstetrics and Gynecology, Huddinge University Hospital,
141 86 Huddinge, Sweden

J. Wilpshaar
Department of Hematology, Leiden University Medical Center, K6-31,
2300 RC Leiden, The Netherlands

A. Wodnar-Filipowicz
Laboratory of Experimental Hematology, Research Department,
Hebelstrasse 20, 4031 Basel, Switzerland

E. Wunder
Stem Cell Laboratory, Institut de Reserche en Hematologie-Transfusion,
Hopital de Hasenrain, Avenue d'Altkirch, 68051 Mulhouse, France

1 Plasticity of Stem Cells

U. Kapp, R. Mertelsmann

1.1 Introduction

The interest in hematopoietic stem cells has increased over the past decades, because they are used as targets for gene transfer in gene therapy as well as serving as the source for bone marrow transplantation (BMT). BMT has become a well-established therapy, and its application has been extended to transplantation of unrelated individuals as well as older individuals. This is due to the use of less toxic or non-myeloablative conditioning protocols, the shortening of the duration of aplasia by the application of growth factors, and effective therapy of infectious disease or sepsis by potent antibiotic regimens. For many patients suffering from leukemia or myelodysplastic syndrome, BMT is the only therapeutic approach that potentially cures their disease. Currently, sources of hematopoietic stem cells for BMT are bone marrow (BM), mobilized peripheral blood stem cells (PBSC) and cord blood (CB). In particular, the use of PBSC has been increasing. When compared with

BM, the collection of PBSC is of lower risk for the donor because there is no need for general anesthesia, the engraftment of stem cells is faster, and the number of collected immature CD34[+] hematopoietic cells is higher. In the autologous setting, immature hematopoietic cells can be purified, e.g., by CD34-selection, to prevent contamination of the graft by tumor cells. However, do we know exactly which cells we have to collect? Do we know enough about *human* hematopoietic stem cells?

The hematopoietic system is organized as a hierarchy that is comprised of frequent, short-lived mature cells of different hematopoietic lineages (Ogawa 1993; Orlic and Bodine 1994; Morrison et al. 1995). These develop from more immature and less frequent progenitor cells that differentiate from a very rare population of pluripotent, long-lived hematopoietic stem cells (HSC). HSC are characterized by their capacity to self-renew and differentiate into different hematopoietic lineages. The only conclusive assay to study HSC is to follow their potential to repopulate lethally irradiated recipients. Therefore, it is difficult to study *human* stem cells. Most of our knowledge about stem cells has been gained from the murine system or in vitro studies. To overcome this problem J.E. Dick and coworkers developed an alternative repopulation assay for human hematopoietic stem cells in immunodeficient NOD/SCID mice (Lapidot et al. 1992; Dick et al. 1997), which shall be discussed below. However, little is known about human HSC and their relations to other pluripotent stem cells such as mesenchymal stem cells or neural stem cells, which have recently been found in the BM. Whether or not BM is a source of a common human pluripotent stem cell for tissues derived from different germ layers remains an open question.

1.2 Methods to Assay Human HSC

1.2.1 In Vitro Assays

In vitro assays for hematopoietic progenitor cells include clonogenic assays in semisolid media, which detect colony-forming cells (CFC). CFC are committed progenitor cells for the myeloid lineages and have a limited proliferative potential (Eaves et al. 1992).

More immature hematopoietic cells can be grown in long-term cultures (LTC) on stroma for up to 60 days and, more recently, in extended

LTC for up to 100 days (Eaves et al. 1992; Hao et al. 1995). Cells capable of initiating hematopoiesis on stroma in those cultures are called LTC-initiating cells (LTC-IC).

1.2.2 NOD/SCID Repopulating Assay

The only conclusive way to assay HSC is to reconstitute the entire hematopoietic system of a conditioned recipient. These experiments were restricted to the murine system until alternative repopulation-assays for HSC were developed by xenotransplantation into immunodeficient mice such as beige-nude-X-linked immunodeficient (BNX) (Nolta et al. 1994; Dao et al. 1998) – or non-obese diabetic (NOD)/severe combined immunodeficient (SCID) mice (Lapidot et al. 1992; Dick et al. 1997). Applying these in vivo models, HSC from BM, PBSC or CB are injected intravenously into sublethally irradiated mice. After an observation period of between 4 weeks and 6 months the mice are killed, and their BM or other tissues are examined for engraftment with human cells by clonogenic assays for human progenitor cells, flow cytometry with a human-specific antibody against CD45, or Southern blot analysis with an alpha-satellite probe specific for human chromosome 17. Especially high levels of human engraftment have been observed in NOD/SCID mice, even without any growth factor treatment (Vormoor et al. 1994). Human cells capable of repopulating NOD/SCID mice are called SCID-repopulating cells (SRC).

1.2.2.1 Phenotype of SRC
In order to answer the question of whether repopulating SRC might be different from any primitive cells that can be assayed in vitro, many efforts have been undertaken to functionally and phenotypically characterize SRC. In order to evaluate the phenotype of SRC, selected cell populations have been transplanted into NOD/SCID mice (Bhatia et al. 1997). Primitive human HSC were purified using a stringent two-step method. In the first step, lineage positive cells were depleted by immunomagnetic separation; secondly, the cells were purified by fluorescence-activated cell sorting (FACS) into $CD34^+CD38^+Lin^-$ and $CD34^+CD38^-Lin^-$ cell-fractions. In these experiments, SRC were found exclusively in the very primitive $CD34^+CD38^-Lin^-$ population. Since

LTC-IC can also be grown from CD34+CD38+ cells, these data suggest that SRC might be more primitive in the hierarchy of the hematopoietic system. Through limiting dilution analysis using Poisson statistics, the frequency of SRC was estimated to be 1 in 617 CD34+CD38− cells. Recently M. Bhatia and co-workers were able to demonstrate that SRC are also present in the CD34−CD38−Lin− cell compartment (Bhatia et al. 1998). Limiting dilution analysis showed a much lower frequency of 1 SRC among 125,000 cells within that cell fraction. However, the CD34−CD38− cells have been shown to differentiate into CD34+ cells in vitro as well as in vivo, which would suggest that CD34−CD38− HSC could be more primitive than CD34+ HSC.

1.2.2.2 Multilineage Differentiation of SRC

After transplantation of early human CD34+/CD38− HSC into NOD/SCID mice, multilineage differentiation could be observed (Bhatia et al. 1997). As shown by FACS analysis, SRC are capable of extensive proliferation and multilineage differentiation in NOD/SCID mice. Losing the expression of CD34, the cells differentiate into more mature CD19+ B cells that also partially express CD20 and surface immunoglobulin M (IgM), and into CD33-expressing myeloid cells, which also partially express CD14 or CD15. However, after transplantation of purified CD34+CD38−, no differentiation into T lymphocytes could be observed.

1.2.2.3 Influence of Growth Factors and Accessory Cells on the Differentiation and Engraftment of SRC

The Differentiation of SRC Can be Modulated by Growth Factor Treatment. Transplantation of SRC induces multilineage engraftment in NOD/SCID mice. In untreated mice the human graft regularly consists of a majority of immature B lymphocytes and CD33+ myeloid cells. To investigate whether the NOD/SCID mouse model could be used as an in vivo system to study the effect of growth factors on human cells, we tested the in vivo effects of FLT3 ligand (FL) and interleukin 7 (IL7) (Kapp et al. 1998). The mice were treated with FL, IL7 or a combination of both every 2nd day by intraperitoneal injection. The phenotype of the human cells harvested from the BM of these mice were characterized by flow cytometry. The analysis of control mice without cytokine treatment confirmed that the NOD/SCID mouse seems to supply a perfect microenvironment for proliferation and differentiation of B cells. In this group, 77% of the human CD45-positive cells expressed CD19. Cells derived from mice, which were treated with IL7 alone showed the same lineage distribution, with 73% CD19-positive B cells among CD45$^+$ human cells. Treatment of the mice with FL changed the lineage distribution of the resulting human cells towards the myeloid lineage. This effect was enhanced by combining FL with IL7. Compared with 10% of CD45$^+$ in untreated mice, 44% of the human cells were committed to the myeloid lineage after treatment with FL and IL7. FL and FL in combination with IL7 seem to impede the B-cell differentiation of immature human SCID mouse engrafting cells. The effect on lineage distribution was similar when highly purified cells were transplanted, suggesting a direct effect on the engrafting cells. These data demonstrate that the differentiation of the human graft in NOD/SCID mice can be modulated by administration of cytokines, providing an important tool to evaluate the in vivo effects of human cytokines.

Growth Factor Treatment and/or Co-transplantation of Non-engrafting Accessory Cells is Necessary to Induce Engraftment by Limiting Doses of CD34$^+$/CD38$^-$ Cells. Little is known about cell types and factors that play a role in the engraftment process. To clarify whether cytokines or cell-to-cell contact might be involved in engraftment, limiting doses (500–1000) of engrafting CD34$^+$CD38$^-$Lin$^-$ cells were transplanted into NOD/SCID mice that were either additionally transplanted with non-engrafting accessory cells (AC) and/or growth factor-treated (Bonnet et al. 1999). The co-transplanted AC population were non-engrafting CD34$^+$CD38$^+$Lin$^-$ or CD34$^-$Lin$^+$ cells. The cytokines that were alternatively or additionally injected were SCF, interleukin 3 (IL3) and granulocyte-macrophage colony-stimulating factor (GM-CSF). Mice that were transplanted with limiting doses of engrafting cells required the co-transplantation of AC or short-term in vivo treatment with cytokines. If higher doses of engrafting cells (>5000) were transplanted, no further treatment of the mice was necessary. Interestingly, cytokine treatment was only effective during the first 10 days. No engraftment was detected in mice which were transplanted with limiting doses of engrafting cells, if cytokine treatment was delayed until 10 days after transplant. These data show that the engraftment process requires pluripotent stem cells as well as growth factors or AC. Which AC are most effective remains to be specified in further studies.

1.3 Gene Transfer into Primitive HSC

The study of the developmental potential of early hematopoietic stem cells or stem cells in general could be facilitated by the introduction of genes into human stem cells. Furthermore, HSC would be ideal targets for gene transfer in gene therapy because of their long life span.

1.3.1 Gene Transfer into Human HSC in a Clinical Trial

Ex vivo culture of BM or PBSC before autologous transplantation can be used to eliminate malignant cells such as Philadelphia chromosome-positive CML cells from non-malignant stem cells. In order to clarify the in vivo activity of non-malignant and malignant stem cells derived from such transplants, gene marking was performed during 5-day ex vivo culture of autologous stem cells by C. v. Kalle and co-workers in a CML retroviral gene marking trial. The protocol uses two purging strategies: early apheresis after chemotherapy and extended 5-day serum-free ex vivo culture with PG13/LN retroviral gene marking. Retroviral gene transfer into normal and malignant repopulating cells was assessed. After myeloablative chemotherapy, ex vivo cultured (10% of cells) and non-cultured autologous PBSC (90% of cells) were co-transplanted. Five-day extended culture in serum-free vector containing medium allowed an efficient expansion (1.3- to 5.1-fold) and transduction efficiency (12% to 13.5%) of LTC-IC from pre-transplant PBSC ($n=3$). Gene transfer could be observed in up to 1% of human peripheral blood mononuclear cells early (day 24) after transplantation. With the help of ligation-mediated polymerase chain reaction (PCR) and sequencing of the PCR products, retroviral integration sites were successfully used to monitor the clonality of marked stem cells in humans. The first data from C. von Kalle and co-workers show successful gene marking currently at >600 days and demonstrate the activity of different stem cell-clones at different time points (von Kalle, personal communication). These observations confirm that cells capable of long-term hematopoiesis can divide and survive during the 5 day ex vivo transduction culture.

1.3.2 Gene Transfer into SRC

Murine stem cells can be transduced efficiently by retroviral gene transfer. In contrast, gene marking is much less efficient in stem cells derived from large mammals such as dogs, primates and humans, as shown in clinical gene transfer trials (van Beusechem et al. 1992; Carter et al. 1992; Bodine et al. 1993). Gene transfer protocols applied in human clinical trials are being optimized using in vitro CFC and LTC-IC assays. However, these progenitors might not have the same properties as repopulating stem cells. To find an assay system that most closely reflects gene marking of repopulating stem cells, Dick and co-workers compared gene transfer efficiency in CFC, LTC-IC and SRC. While high fractions of CFC and LTC-IC (up to 75%) were transduced, SRC were only rarely transduced (Larochelle et al. 1996). Also, after optimizing the method e.g., by the use of fibronectin, purification of CD34+ cells or infection of larger numbers of SRC, SRC with a high proliferative capacity were only occasionally transduced. The transduction efficiency into very primitive cells seems to be low, which is consistent with results from studies in primates, dogs, and human clinical trials. Thus, the NOD/SCID mouse model seems to be an appropriate model to optimize gene transfer protocols and to estimate the gene transfer efficiency into very primitive human repopulating cells, as opposed to progenitors such as CFC or LTC-IC. Again, the data from the NOD/SCID model indicate that SRC are different from CFC and LTC-IC. SRC seem to be the most primitive human hematopoietic cells that can be assayed in an experimental system.

1.4 Cells Different from Hematopoietic Stem Cells in Human BM or Stem Cell Transplantation

1.4.1 Human Cells Different from Hematopoietic Cells Engraft NOD/SCID Mice

After xenotransplantation of separated CD34$^+$ cells isolated from human PBSC or CB into NOD/SCID mice, Henschler and co-workers demonstrated engraftment of hematopoietic and non-hematopoietic cells. In the BM of xenotransplanted mice, 0.1–1% of the cells expressed human von-Willebrand factor (vWF), which is characteristic of endothelial cells (Junghahn et al. 1998). In chimeric BM cultures grown from engrafted mice and supplemented with rh-VEGF, growth of human endothelial cells could be induced. The endothelial phenotype was demonstrated by immunohistochemical staining with human-specific antibodies against vWF, CD31 and VEGF-receptor-2 (KDR). It is unclear at this point whether these cells are derived from co-transplanted mature endothelial cells, which have been demonstrated in CB or in PB previously, or from more immature endothelial precursor cells.

1.4.2 Pluripotent Cells in the Bone Marrow

Types of pluripotent cells that have been demonstrated in BM recently include: (1) hematopoietic stem cells, (2) liver-derived oval cells, (3) mesenchymal stem cells (MSC), (4) neural cells, and (5) myogenic progenitors.

1.4.2.1 BM-Derived Cells Differentiate into Hepatocytes
When hepatocytes are prevented from proliferation in response to liver damage, hepatic oval cells get induced to proliferate and may function as progenitors for two types of epithelial cells in the liver: hepatocytes and bile duct cells. Interestingly, oval cells express CD34, Thy-1, c-kit, and FLT3-R like HSC. Petersen et al. (1999) investigated whether oval cells and other liver cells can derive from a cell population that is originated from or associated with extrahepatic tissue such as the BM. Three different approaches have been applied: In a first experiment, female rats were lethally irradiated and repopulated with a BM trans-

plant from syngeneic male animals. Donor cells were detected by DNA probes specific for the Y chromosome. In a second set of experiments, BM was transplanted from dipeptidyl peptidase IV (DPP IV)-positive male rats into DPP IV-negative syngeneic females. Donor cells were detected by cytochemically assaying the activity of the DPP IV enzyme. As a third approach, liver transplantation was performed with livers from Brown Norway donor rats into Lewis rats. In contrast to Lewis rats, the Brown Norway rats do not express the L21- antigen, and every L21-6 antigen-expressing cell in transplanted Brown Norway livers must be derived from an extrahepatic source. After successful transplantation, the animals were exposed to 2-acetylaminofluorene (2-AAF) to inhibit hepatocyte proliferation, and hepatic injury was performed by partial hepatectomy or by administration of carbon tetrachloride (CCl4). It could be demonstrated in all three experiments that oval cells must be partially derived from an extrahepatic source, which in two of the approaches was shown to be BM. These data suggest that a cell associated with the BM may have the potential to act under certain physiopathological conditions as a precursor cell for different types of liver cells.

1.4.2.2 Mesenchymal Stem Cells

Other pluripotent precursor cells associated with the BM that have raised the interest of researchers working on tissue repair and engineering are mesenchymal stem cells (MSC) (Fig. 1). In the mid-1970s, Friedenstein isolated fibroblast-like cells from samples of whole BM (Friedenstein et al. 1976; Prockop 1997). He placed unselected BM cells into plastic culture dishes, and after 4 h cells that were non adherent were removed. The most striking finding was that these tightly adherent cells, which began to proliferate rapidly after 2–4 days, had the ability to differentiate into colonies that resembled little foci of bone or cartilage. Since then, a number of studies have shown that these cells, termed mesenchymal stem cells (MSC), can be differentiated into osteoblasts, chondroblasts, adipocytes and myoblasts (Prockop 1997; Pittenger et al. 1999). The osteogenic differentiation can be induced by addition of dexamethasone, β-glycerol phosphate and ascorbate. Treatment with 1-methyl-3-isobutylxanthine, dexamethasone, insulin and indomethacin promotes adipogenic differentiation. Induction of myoblast differentiation follows after addition of 5-azacytidine. Finally chondrogenic differ-

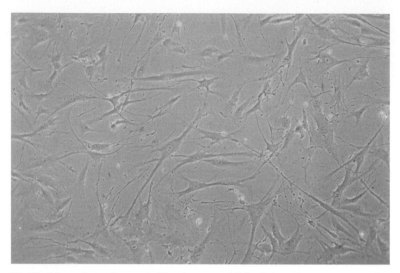

Fig. 1. Mesenchymal stem cell culture established from bone marrow after density fractionation

entiation can be observed after treatment with transforming growth factor (TGF)-β. The mechanisms leading to selective differentiation are poorly understood. In a recent study, Pittenger et al. (1999) demonstrated that individual MSC can be isolated and expanded to colonies that retain their multilineage potential. These expanded MSC are uniformly positive for CD29, CD44, CD71, CD90, CD106, CD120a, and CD124. They do not express markers of the hematopoietic lineage such as CD14, CD34, and CD45. Further characterization of MSC, especially the investigation of their in vivo capacities, may provide the potential for new therapeutic approaches for the restoration of diseased or damaged tissue. Transplantation of MSC could potentially be used therapeutically to correct genetic disorders of bone, cartilage, and muscle.

In a recent study, allogeneic BM transplantation was performed in three children suffering from osteogenesis imperfecta (OI), a genetic disorder leading to osteopenia, shortened stature, severe bony deformities, and multiple fractures (Horwitz et al. 1999). This phenotype is due to production of defective type I collagen in osteoblasts. Three months after allogeneic BM transplantation, specimens of trabecular bone showed histologic changes that indicated new dense bone formation. In all patients, the total body mineral content increased, which was associated with increases in growth velocity and reduced frequencies of bone fractures. Surprisingly, these significant effects were correlated with the presence of only 1.5–2% donor mesenchymal cells when evaluated by fluorescence in situ hybridization to detect the Y chromosome or by DNA polymorphism analysis on day 80–100. How the presence of only few donor-mesenchymal cells can lead to such improvement could be explained by observations in asymptomatic mosaic parents, where the ratio of mutated to normal alleles in some tissues approaches the value seen in their affected children. The severity of the disease is dependent on the relative balance between synthesis of mutated and normal polypeptide chains. Low levels of mesenchymal cell engraftment might be enough to shift this balance to a level that is sufficient to convert a severe phenotype to a less severe one. Furthermore, a small number of normal osteoblasts could have an impact on the osteogenic microenvironment. Another explanation could be a short-lived engraftment of mesenchymal cells. The percentage of donor osteoblasts might have been greater early after engraftment, which was not evaluated in this study. A temporarily higher number of donor osteoblasts may have produced an increased amount of normal collagen fibers, providing a matrix for the deposition of mineral. It will be important to further investigate whether early improvements in bone structure and function are due to transplanted self-renewing stem cells or to progenitors with only limited proliferative capacity. However, this study on allogeneic BMT in OI patients indicates that BMT may serve as a therapeutic approach in diseases of mesenchymal progenitors such as OI and, possibly, even muscular disease.

Muscle regeneration by BM-derived myogenic progenitors has been demonstrated in the mouse system (Ferrari et al. 1998). Unfractionated BM cells from the C57/MlacZ transgenic mouse line, in which a lacZ gene encoding a nuclear β-galactosidase (β-Gal) is under the control of a muscle-specific myosin promoter, were injected into the tibialis anterior (TA) muscle of immunodeficient scid/bg mice following chemically induced muscle damage and regeneration. Two weeks after injection of total BM, histochemical staining of the TA muscles revealed fibers containing β-Gal-positive aligned nuclei. In a second experiment, irradiated scid/bg mice were transplanted with genetically marked BM cells from the C57/MlacZ line. Five weeks after transplantation, muscle regeneration was induced in both TA muscles of the repopulated mice. Histochemical analysis of the muscle showed regenerating fibers containing β-Gal-positive nuclei in five out of six reconstituted mice. These data indicate the existence of BM-derived myogenic progenitors which can migrate into damaged muscle and participate in the regeneration process. It is possible that these cells originated from multipotential MSC. Thus BMT could be a means to correct genetic muscle disease.

As discussed previously, MSC can be differentiated into different cell types of mesenchymal origin under certain in vitro conditions. A recent study investigating the in vivo growth of MSC in the murine system demonstrated the in vivo potential of MSC to differentiate into astrocytes and possibly neurons (Kopen et al. 1999). MSC were isolated from murine BM and labeled with BrdU or with bis-benzimide, a fluorescent DNA-binding dye. The labeled cells were injected into the lateral ventricle of neonatal mice. The MSC could be shown to migrate throughout the forebrain and cerebellum about 12 days after transplantation, which is consistent with ongoing developmental processes occurring in early postnatal life. MSC within the striatum and the molecular layer of the hippocampus expressed glial fibrillary acidic protein, indicating differentiation into astrocytes. The population of neuron-rich regions by MSC and the expression of neurofilaments by the donor cells might suggest differentiation of MSC into neurons. These data suggest that MSC mimic the behavior of neural progenitor cells. This study demonstrates for the first time that MSC might differentiate into cells which are supposed to originate from a separate developmental layer.

1.5 Non-hematopoietic Progenitor Cells Induce Hematopoiesis

1.5.1 Neural Progenitor Cells Can Differentiate into Hematopoietic Cells

Bjornson et al. recently claimed the existence of common neurohematopoietic stem cells and entitled his study "Turning brain into blood" (Bjornson et al. 1999). The investigators isolated and clonally propagated neural stem cells (NSC) from mouse brain derived from ROSA26 mice. ROSA26 mice were chosen as the source of donor tissue, because they are transgenic for LacZ encoding the enzyme β-Gal. The ROSA26 NSC were intravenously injected into sublethally irradiated Balb/c mice. After 5–12 months, PCR amplification of the LacZ gene revealed a strong signal in the spleens of mice transplanted with embryonic, adult or clonally expanded adult NSC. BM cells of engrafted mice were plated in methylcellulose in the presence of defined hematopoietic cytokines. About 14 days after plating, colonies of myeloid and B-lymphoid hematopoietic progenitor cells were found to express β-galactosidase. Lymphoid and myeloid engraftment from ROSA26-derived NSC was confirmed by flow cytometric analysis. This study strikingly shows that hematopoiesis can be reconstituted by NSC that are derived from a separate germ layer.

1.5.2 Muscle Progenitor Cells Can Induce Hematopoiesis

In addition to NSC, muscle progenitor cells, also called satellite cells, were found to exhibit hematopoietic potential (Jackson et al. 1999). Satellite cells are progenitors of muscle cells that get activated after injury. Murine Ly-5.1 satellite cells were isolated from adult skeletal muscle by enzymatic digestion and in vitro culture. The cells were harvested after 5 days and injected into lethally irradiated Ly-5.2 recipients. All of them were found to be repopulated with hematopoietic cells or all major blood lineages. Total peripheral blood contribution from satellite cells (Ly-5.1) ranged from 20–80%. These data indicate that an already committed progenitor cell might get reprogrammed and regain its stem cell potential.

1.6 Plasticity of Stem Cells

All the studies discussed above completely change our view of stem cell biology. Our theory of stem cells is based on the hypothesis of a hierarchy with totipotent embryonic stem cells on the top, which differentiate into more specialized progenitor cells belonging to the separate dermal layers. These progenitors diverge into different, more mature cells which loose their capacity to self-renew and differentiate. Re-

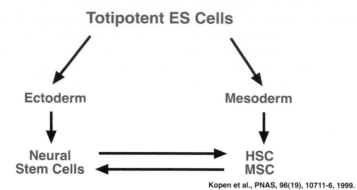

Bjornson et al.: "Turning Brain into Blood..."
Science, 283, 534-537, 1999.

Kopen et al., PNAS, 96(19), 10711-6, 1999.

Fig. 2. Cells from different primary germ layers can differentiate into each other. *ES Cells*, embryonic stem cells; *HSC*, hematopoietic stem cells; *MSC*, mesenchymal stem cells. (After Bjornson et al., Kopen et al.)

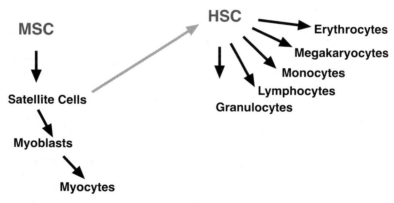

Fig. 3. Progenitor cells can regain stem cell potential

After treating him so cruelly, the soldiers loaded Jesus down with a heavy wooden cross. They led him through the streets of Jerusalem towards Mount Calvary. Here he would be crucified.

Many people followed Jesus. Lots of women were weeping and crying to see how he was suffering, but Jesus comforted them with kind words.

*A*t Mount Calvary,
the soldiers nailed Jesus
to the cross.

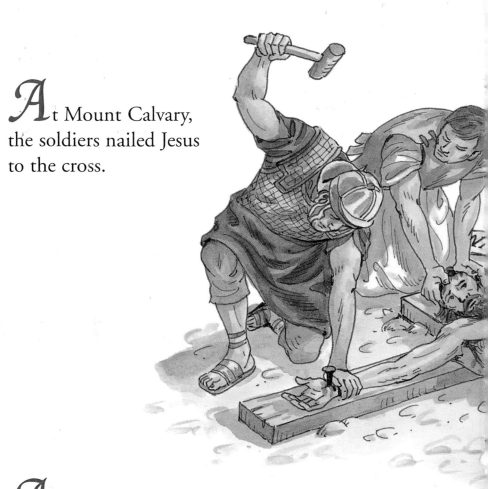

*A*bove the cross, they nailed a sign. It said:
'Jesus of Nazareth, King of the Jews'.

*T*wo thieves were crucified
alongside Jesus.
'Aren't you the Son of
God?' one asked him.
'Why don't you save
yourself, and us, too?'
'We are rightly punished,'
said the other thief.
'This man has done
nothing wrong.'

*T*he second thief turned
his head towards Jesus.
'Lord,' he said, 'remember
me when you come to
your kingdom.'
'I promise,' said Jesus.
'Today, you will be
with me in Paradise.'

*J*esus suffered for many hours on the cross. The sky became quite dark and the sun was blotted out by a huge shadow. At last, Jesus gave a loud cry, 'Father, into your hands I place my spirit.' At that moment, the ground began to tremble and rocks split apart. People ran away in fright. 'Truly,' said one soldier, 'he was the Son of God.'

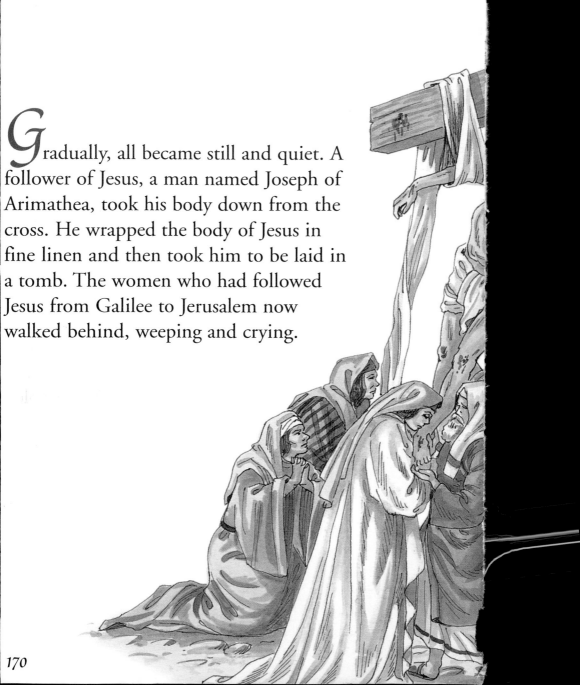

Gradually, all became still and quiet. A follower of Jesus, a man named Joseph of Arimathea, took his body down from the cross. He wrapped the body of Jesus in fine linen and then took him to be laid in a tomb. The women who had followed Jesus from Galilee to Jerusalem now walked behind, weeping and crying.

When Jesus had been laid inside, the entrance to the tomb was sealed with a large stone. The Pharisees feared that the disciples might take away his body and then say he had risen from the dead, as Jesus had said he would. So, for three days, men were on guard outside the tomb.

On the third day after the
death of Jesus, two women came
to his tomb. They saw that the
great stone had been rolled away.
Bravely, they stepped inside and
there was an angel, gleaming
white. 'Jesus is not here,' said the
angel. 'He is risen! Now he has
gone into Galilee. Go and tell
his disciples!' All that the angel
said was true. Jesus had risen
from the dead.

Jesus stayed with his disciples for forty days. Then he led them to the Mount of Olives. 'Soon,' he said, 'God will send you his Holy Spirit. You will be able to speak different languages so that you can carry on with my work all over the world. I will be with you always.' As Jesus raised his hands to bless them, he was lifted up into heaven.

cently, more and more evidence has arisen which suggests that already specialized tissue-specific stem cells such as NSC or MSC can serve as stem cells for cells from a different dermal origin (Fig. 2.). Furthermore, committed progenitor cells such as satellite cells of the muscle may get reprogrammed and regain the potential to function as hematopoietic stem cells (Fig. 3). Lineage defined progenitor cells in adult tissues seem to be more plastic than hitherto thought. In contrast to our earlier concepts, normal, non-malignant cells under certain conditions can dedifferentiate and get reprogrammed. The idea of a hierarchy gets replaced by the new paradigm of stem cell plasticity.

Acknowledgements. The authors thank Dr. M. Follo for carefully reading the manuscript.

References

Bhatia M, Bonnet D, Murdoch B, Gan OI, Dick JE (1998) A newly discovered class of human hematopoietic cells with SCID-repopulating activity (see comments). Nat Med 4(9):1038–1045

Bhatia M, Wang JCY, Kapp U, Bonnet D, Dick JE (1997) Purification of primitive human hematopoietic cells capable of repopulating immune-deficient mice. Proc Natl Acad Sci USA 94(10):5320–5325

Bjornson CR, Rietze RL, Reynolds BA, Magli MC, Vescovi AL (1999) Turning brain into blood: a hematopoietic fate adopted by adult neural stem cells in vivo (see comments). Science 283(5401):534–537

Bodine DM, Moritz T, Donahue RE, Luskey BD, Kessler SW, Martin DI, Orkin SH, Nienhuis AW, Williams DA (1993) Long-term in vivo expression of a murine adenosine deaminase gene in rhesus monkey hematopoietic cells of multiple lineages after retroviral mediated gene transfer into CD34+ bone marrow cells. Blood 82(7):1975–1980

Bonnet D, Bhatia M, Wang JC, Kapp U, Dick JE (1999) Cytokine treatment or accessory cells are required to initiate engraftment of purified primitive human hematopoietic cells transplanted at limiting doses into NOD/SCID mice. Bone Marrow Transplant 23(3):203–209

Carter RF, Abrams-Ogg AC, Dick JE, Kruth SA, Valli VE, Kamel-Reid S, Dube ID (1992) Autologous transplantation of canine long-term marrow culture cells genetically marked by retroviral vectors. Blood 79(2):356–364

Dao MA, Nolta JA, Hanley MB, Kohn DB (1998) Use of the bnx/hu xenograft model of human hematopoiesis to optimize methods for retroviral-mediated stem cell transduction (review). Sustained human hematopoiesis in immunodeficient mice by cotransplantation of marrow stroma expressing human interleukin-3: analysis of gene transduction of long-lived progenitors. Int J Mol Med 1(1):257–264

Dick JE, Bhatia M, Gan O, Kapp U, Wang JC (1997) Assay of human stem cells by repopulation of NOD/SCID mice. Stem Cells 15 [Suppl 1]:199–203; discussion 204–207

Eaves CJ, Sutherland HJ, Udomsakdi C, Lansdorp PM, Szilvassy SJ, Fraser CC, Humphries RK, Barnett MJ, Phillips GL, Eaves AC (1992) The human hematopoietic stem cell in vitro and in vivo (see comments). Blood Cells 18(2):301–307

Ferrari G, Cusella-De Angelis G, Coletta M, Paolucci E, Stornaiuolo A, Cossu G, Mavilio F (1998) Muscle regeneration by bone marrow-derived myogenic progenitors. Science 279(5356):1528–1530

Friedenstein AJ, Gorskaja JF, Kulagina NN (1976) Fibroblast precursors in normal and irradiated mouse hematopoietic organs. Exp Hematol 4(5):267–274

Hao QL, Shah AJ, Thiemann FT, Smogorzewska EM, Crooks GM (1995) A functional comparison of CD34+ CD38- cells in cord blood and bone marrow. Blood 86(10):3745–3753

Horwitz EM, Prockop DJ, Fitzpatrick LA, Koo WW, Gordon PL, Neel M, Sussman M, Orchard P, Marx JC, Pyeritz RE, Brenner MK (1999) Transplantability and therapeutic effects of bone marrow-derived mesenchymal cells in children with osteogenesis imperfecta (see comments). Nat Med 5(3):309–313

Jackson KA, Mi T, Goodell MA (1999) Hematopoietic stem cells in adult skeletal muscle. Abstract (LBA035), International Society of Experimental Hematology, annual scientific meeting, Monte Carlo

Junghahn I, Goan SR, Fichtner I, Becker M, Möbest D, Henschler R (1998) Circulating endothelial cells from adult mobilized as well as from placental blood engraft in NOD/SCID mice. Abstract 2897, American Society of Hematology, annual scientific meeting, Miami

Kapp U, Bhatia M, Bonnet D, Murdoch B, Dick JE (1998) Treatment of nonobese diabetic (NOD)/Severe-combined immunodeficient mice (SCID) with flt3 ligand and interleukin-7 impairs the B-lineage commitment of repopulating cells after transplantation of human hematopoietic cells. Blood 92(6):2024–2031

Kopen GC, Prockop DJ, Phinney DG (1999) Marrow stromal cells migrate throughout forebrain and cerebellum, and they differentiate into astrocytes after injection into neonatal mouse brains. Proc Natl Acad Sci USA 96(19):10711–10716

Lapidot T, Pflumio F, Doedens M, Murdoch B, Williams DE, Dick JE (1992) Cytokine stimulation of multilineage hematopoiesis from immature human cells engrafted in SCID mice. Science 255(5048):1137–1141

Larochelle A, Vormoor J, Hanenberg H, Wang JC, Bhatia M, Lapidot T, Moritz T, Murdoch B, Xiao XL, Kato I,Williams DA, Dick JE (1996) Identification of primitive human hematopoietic cells capable of repopulating NOD/SCID mouse bone marrow: implications for gene therapy. Nat Med 2(12):1329–1337

Morrison SJ, Uchida N, Weissman IL (1995) The biology of hematopoietic stem cells. Annu Rev Cell Dev Biol 11:35–71

Nolta JA, Hanley MB, Kohn DB (1994) Sustained human hematopoiesis in immunodeficient mice by cotransplantation of marrow stroma expressing human interleukin-3: analysis of gene transduction of long-lived progenitors. Blood 83(10):3041–3051

Ogawa M (1993) Differentiation and proliferation of hematopoietic stem cells. Blood 81(11):2844–2853

Orlic D, Bodine DM (1994) What defines a pluripotent hematopoietic stem cell (PHSC): will the real PHSC please stand up! (Editorial). Blood 84(12):3991–3994

Petersen BE, Bowen WC, Patrene KD, Mars WM, Sullivan AK, Murase N, Boggs SS, Greenberger JS, Goff JP (1999) Bone marrow as a potential source of hepatic oval cells. Science 284(5417):1168–1170

Pittenger MF, Mackay AM, Beck SC, Jaiswal RK, Douglas R, Mosca JD, Moorman MA, Simonetti DW, Craig S, Marshak DR (1999) Multilineage potential of adult human mesenchymal stem cells. Science 284(5411):143–147

Prockop DJ (1997) Marrow stromal cells as stem cells for nonhematopoietic tissues. Science 276(5309):71–74

van Beusechem VW, Kukler A, Heidt PJ, Valerio D (1992) Long-term expression of human adenosine deaminase in rhesus monkeys transplanted with retrovirus-infected bone-marrow cells. Proc Natl Acad Sci USA 89(16):7640–7644

Vormoor J, Lapidot T, Pflumio F, Risdon G, Patterson B, Broxmeyer HE, Dick JE (1994) Immature human cord blood progenitors engraft and proliferate to high levels in severe combined immunodeficient mice. Blood 83(9):2489–2497

2 Lentiviral Vectors for the Genetic Modification of Hematopoietic Stem Cells

D. Trono

Lentiviral vectors, owing to their ability to deliver transgenes in tissues that long appeared irremediably refractory to stable genetic manipulation, open exciting perspectives for the genetic treatment of a wide array of hereditary as well as acquired disorders, in particular of the lympho-hematopoietic system.

2.1 From Oncoretroviral to Lentiviral Vectors

Retroviral vectors, classically derived from oncoretroviruses such as the murine leukemia virus, present three highly appealing characteristics for a gene delivery system. First, they have a quite large cloning capacity, close to 10 kb. Second, they integrate their cargo into the chromosomes of target cells, a likely prerequisite for long-term expression. Third, they do not transfer virus-derived coding sequence, avoiding transduced cells being recognized and destroyed by vector-specific cytotoxic T lympho-

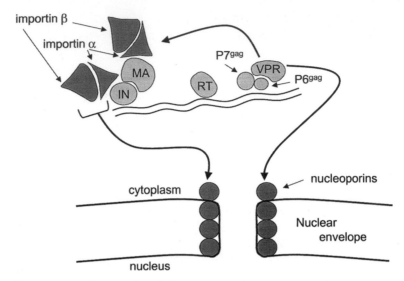

Fig. 1. Nuclear import of the HIV-1 preintegration complex (PIC). Nuclear localization signals (NLS) on the viral matrix (*MA*) and integrase (*IN*) are recognized by the NLS receptor importin α. Importin β is then recruited, allowing docking on the cytoplasmic edge of the nucleopore through the recognition of nucleoporins by the importin α-importin β complex. Virion protein R (*VPR*), recruited in the PIC by p6^{gag}, apparently strengthens the interaction between NLS-bearing proteins and importin α, and can also interact directly with nucleoporins. *Wavy double line,* viral genome

cytes. In spite of these impressive assets, however, oncoretroviral vectors suffer from a striking limitation: they cannot transfer genes into cells that do not divide within a few hours of being inoculated (Roe et al. 1993; Lewis and Emerman 1994). Yet most of the potential targets of gene therapy are cells that rarely if ever proliferate, be they neurons, hepatocytes, myocytes, or hematopoietic stem cells (HSC).

In contrast to oncoretroviruses, lentiviruses such as the human immunodeficiency virus type 1 (HIV-1) can replicate in non-mitotic cells because their so-called preintegration complex, a macromolecular structure comprising the viral genome, a few structural proteins and the enzymes responsible for reverse transcription and integration, can hijack the cell nuclear import machinery (Bukrinsky et al. 1992; Lewis et al.

1992; Gallay et al. 1996). The vector genome is thereby docked at the nuclear envelope and transported through the nucleopore, allowing integration in cells such as lowly activated T lymphocytes and terminally differentiated macrophages. This property reflects the karyophilic properties of several viral constituents that play both additive and redundant roles: for HIV-1, matrix (MA), virion protein R (VPR), and integrase (IN) (Bukrinsky et al. 1993; Heinzinger et al. 1994; von Schwedler et al. 1994; Gallay et al. 1997) (Fig. 1). The complexity of their action within the highly organized lentiviral preintegration complex probably explains why they could not be transferred easily to oncoretroviral vectors. Faced with this difficulty, the option was taken to develop a vector straight from the best characterized of all lentiviruses, HIV-1. This led to the proof-of-principle that lentiviral vectors can mediate the efficient in vivo delivery, stable integration, and long-term expression of transgenes into non-mitotic cells such as neurons (Naldini et al. 1996). Broad perspectives were opened, although it was clear that significant improvements would have to be made before the clinical use of this gene delivery tool.

2.2 Towards Lentiviral Vectors Acceptable for Clinical Use

To derive a vector from a virus, one needs to discriminate the functions that (1) are necessary for the transfer of the virus genetic cargo, (2) are responsible for its further replication within the target cell, and (3) play the role of virulence factors within the context of an organism. Ideally, only the first set of these functions should be conserved in the vector manufacturing system. When the development of HIV-based lentiviral vectors was initiated, the necessary information was largely available through previous work done either on HIV itself or on oncoretroviral vectors.

Three components are involved in making a transducing lentivector particle: first, the genetic information encoding for the virion packaging elements, both structural and enzymatic; second, a nucleotide sequence that codes for or constitutes the vector genome; and third, a so-called producer cell in which these two sequences are introduced either transiently or stably. Typically, lentiviral vector particles are generated in 293 human embryonic kidney cells (Naldini et al. 1996). In the case of

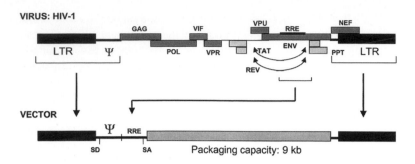

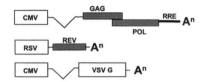

Fig. 2. Lentiviral vector packaging system. *Top*: HIV-1 genome. *Middle*: Structure of a prototypic HIV-1-derived vector, in which only the cis-acting sequences of the viral genome are incorporated: the two long terminal repeats (*LTR*), the packaging signal (ψ), the Rev-responsive element (*RRE*), and the polypurine tract (*PPT*). *SD* and *SA*, splice donor and acceptor sites. *Bottom*: Third-generation HIV-1-derived packaging system

HIV-1-based vectors, the core and enzymatic components of the virion come from HIV-1, while the envelope is derived from a heterologous virus, most often vesicular stomatitis virus due to the high stability and broad tropism of its G protein. The genomic complexity of HIV, where a whole set of genes encodes virulence factors essential for pathogenesis but dispensable for transferring the virus genetic cargo, has allowed the creation of multiply attenuated packaging systems. The latest ("third") generation of these systems comprises only three of the nine genes of HIV-1: *gag*, coding for the virion main structural proteins, *pol*, responsible for the retrovirus-specific enzymes, and *rev*, which encodes a post-transcriptional regulator necessary for efficient *gag* and *pol* expression (Dull et al. 1998) (Fig. 2). The parental virus cannot emerge even through the most acrobatic recombination events, because only 40% of its genome are left in the system.

The vector itself is the only piece of genetic material transferred to the target cells. It typically comprises the transgene cassette flanked by cis-acting elements necessary for its encapsulation, reverse transcription, and integration. As previously done with oncoretroviral vectors, advantage was taken of the gymnastics of reverse transcription to engineer self-inactivating (SIN) HIV-1-derived vectors, which lose the transcriptional capacity of the viral long terminal repeat (LTR) once transferred to target cells Zufferey et al. 1997; Miyoshi et al. 1998). This minimizes the risk of emergence of replication competent recombinants (RCR) and avoids problems linked to promoter interference.

Finally, the biosafety of HIV-based lentiviral vectors was further increased by distributing the sequences encoding its various components over four independent units, to maximize the number of crossovers that would be required to re-create an RCR (Zufferey et al. 1997; Dull et al. 1998). Also, the overlap between vector and helper sequences, the ground for homologous recombination, was reduced to a few tens of nucleotides. And stable packaging cell lines are in the pipeline, which would definitely open the way to the use of lentivectors in the clinics.

Subsequent to the description of the properties of the HIV-1-based gene transfer system, vectors were derived from animal lentiviruses such as simian and feline immunodeficiency viruses and equine infectious anemia virus (Olsen 1998; Poeschla et al. 1998; White et al. 1999). The development and testing of these other systems is less advanced. However, it is likely that they will all have the same general properties as their HIV counterpart, even though they might reveal species-specific limitations that could decrease their utility in some human cells. Biosafety considerations are regularly put forward to justify their use over that of HIV-based vectors. Nevertheless, the genomic complexity of HIV is far greater than that of most other lentiviruses, so that the multiply deleted HIV-based packaging system ends up the farthest away from its parental virus. Second, zoonoses teach us that the pathogenicity of a given organism is largely unpredictable when it is transferred from its normal animal host into humans. Finally, millions of individuals worldwide have been screened for lentivirus-related diseases. Not only has no pathology been associated with massively deleted forms of HIV-1, but on the contrary, well documented cases of long-term clinical non-progression have occurred in patients infected with HIV-1 strains that carry genetic alterations far more subtle than those introduced in the third-generation HIV-1 packaging system.

2.3 Lentiviral Vectors and Human Hematopoietic Stem Cells

The original demonstration that HIV-based vectors can mediate the efficient in vivo delivery and long-term expression of transgenes was provided in the central nervous system, and lentivectors remain a gene delivery system of choice in this tissue (Naldini et al. 1996; Blömer et al. 1997). Lentivectors seem to constitute vehicles of choice for the genetic modification of retinal cells (Miyoshi et al. 1997; Takahashi et al. 1999), and can also efficiently transfer genes in the mammalian cochlea (Han et al. 1999). In contrast, at least in their current form, lentivectors seem rather poorly efficient at delivering genes in vivo into the liver and the muscle. In spite of initially encouraging results (Kafri et al. 1997), adenoviral and adeno-associated viral vectors have a clear-cut edge in these organs.

The lympho-hematopoietic system will likely represent one of the most important targets of lentiviral vectors. Indeed, the development of gene therapy strategies to correct hematopoietic disorders has been hampered by the low efficacy of oncoretroviral vectors at transducing the essentially non-proliferating repopulating human HSC (hHSC). To circumvent this difficulty, HSC are classically treated with cytokines that induce their division and hence make them susceptible to oncoretrovector-mediated transduction. Although this approach is satisfactory in the case of murine HSC, hHSC largely lose their pluripotentiality when induced to proliferate. Solving this problem, several recent studies have demonstrated that VSV G-pseudotyped HIV-based vectors can efficiently transduce human CD34$^+$ HSC in the absence of cytokine stimulation (Sutton et al. 1998; Uchida et al. 1998;Case et al. 1999; Douglas et al. 1999; Miyoshi et al. 1999). Furthermore, lentivector-transduced hHSC are capable of long-term engraftment of non-obese diabetic/severe combined immunodeficient (NOD/SCID) mice (Miyoshi et al. 1999). Finally, bone marrow from these primary recipients can repopulate secondary mice with transduced cells, confirming the lentivector-mediated genetic modification of very primitive hematopoietic precursors, most probably bona fide stem cells (Woods et al., in press).

In a typical experiment, bone marrow (BM) or cord blood (CB) CD34$^+$ cells are exposed to an HIV-based vector in serum-free conditions, in the absence of any cytokine or in the sole presence of megakaryocyte growth and development factor (MGDF) or of TPO. The cycling status of the HSC is not modified by this treatment, more than 98% of the cells remaining in G_0 or G_1. To analyze transgene expression, cells are incubated for a further 96 h in medium supplemented with TPO or MDGF and stem cell factor (SCF). Commonly, close to 30% of the cells express the transgene and are positive for the vector DNA by polymerase chain reaction (PCR). This percentage is maintained when evaluating the transduction rate of cells capable of forming hematopoietic colonies in methylcellulose or of reconstituting hematopoietic lineages in NOD/SCID mice, even though lentiviral vectors seem to transduce HSC in G_1, and not in G_0 (Sutton et al. 1999). Finally, as expected from the persistence of the vector genome through primary and secondary BM transplantation, proviral integration into the host cell chromosome can be demonstrated (Woods et al., in press).

2.4 Lentivector-Mediated Transduction of HSC: Perspectives

Lentiviral vectors offer for the first time the opportunity to alter efficiently and stably the genetic baggage of hHSC, paving the way to a whole new line of investigation. For instance, transgene expression in stem cells is rarely the final goal. Instead, the lineage-specific and regulated production of proteins is most often desired, with requirements that differ greatly according to the nature of the underlying disease. Lentivectors best suited for restoring the function of granulocytes in chronic granulomatous disease will thus likely differ markedly from those achieving the intracellular immunization of T cells and macrophages against HIV infection. Appropriate in vitro and animal models of human diseases will thus need to be utilized to develop and test lentivectors that are tailored for each individual application. Also, the all-important question of transgene silencing will need to be examined. One hope is that lentivectors, because they are derived from viruses that induce life-long infections, are protected from such a phenomenon. In that respect, the long-term follow-up of rats whose brain was injected with HIV-derived vectors expressing reporter genes is encouraging. Future experiments will determine whether persistence of expression is the rule for other transgenes and in other organs as well, in particular in the lympho-hematopoietic system.

The ability to modify the genetic make-up of HSC will also allow one to address important questions about lympho-hematopoiesis, for instance by expressing genes suspected to influence lineage determination. Along the same line, lentivector-mediated genetic manipulations could be attempted to trigger the expansion of HSC without altering their pluripotentiality. In that respect, our recent experiments indicate that cocktails of lentiviral vectors encoding combinations of cell cycle-inducing, anti-senescence and anti-apoptotic gene products can be used to achieve the conditional immortalization of human primary cells such as endothelial cells, myocytes, or keratinocytes (D. Trono, unpublished). The same technique could be used to amplify blood cell progenitors or to obtain BM stromal cell lines that would facilitate the in vitro maintenance and expansion of HSC.

In conclusion, lentiviral vectors offer ground for new and exciting perspectives in the lympho-hematopoietic system, and a previously unexplored basis for the study of hematopoiesis and for the gene therapy of inherited and acquired lympho-hematopoietic disorders via the genetic modification of hHSC.

References

Blömer U et al (1997) Highly efficient and sustained gene transfer in adult neurons with a lentivirus vector. J Virol 71:6641–6649

Bukrinsky M et al (1992) Active nuclear import of human immunodeficiency virus type 1 preintegration complexes. Proc Natl Acad Sci USA 89:6580–6584

Bukrinsky MI et al (1993) A nuclear localization signal within HIV-1 matrix protein that governs infection of non-dividing cells. Nature 365:666–669

Case SS et al (1999) Stable transduction of quiescent CD34$^+$ CD38$^-$ human hematopoietic cells by HIV-1-based lentiviral vectors. Proc Natl Acad Sci USA 96:2988–2993

Douglas J, Kelly P, Evans JT, Garcia JV (1999) Efficient transduction of human lymphocytes and CD34$^+$ cells via human immunodeficiency virus-based gene transfer vectors. Hum Gene Ther 10:935–945

Dull T et al (1998) A third-generation lentivirus vector with a conditional packaging system. J Virol 72:8463–8471

Gallay P, Chin D, Hope TJ, Trono D (1997) HIV-1 infection of nondividing cells mediated through the recognition of integrase by the importin/karyopherin pathway. Proc Natl Acad Sci USA 94:9825–9830

Gallay P, Stitt V, Mundy C, Oettinger M, Trono, D (1996) Role of the karyopherin pathway in human immunodeficiency type 1 nuclear import. J Virol 70:1027–1032

Han JJ et al (1999) Transgene expression in the guinea pig cochlea mediated by a lentivirus-derived gene transfer vector. Hum Gene Ther 10:1867–1873

Heinzinger NK et al (1994) The Vpr protein of human immunodeficiency virus type 1 influences nuclear localization of viral nucleic acids in nondividing host cells. Proc Natl Acad Sci USA 91:7311–7315

Kafri T et al (1997) Sustained expression of genes delivered directly into liver and muscle by lentiviral vectors. Nature Gen 17:314–317

Lewis PF, Emerman M (1994) Passage through mitosis is required for oncoretroviruses but not for the human immunodeficiency virus. J Virol 68:510–516

Lewis PF, Hensel M, Emerman M (1992) Human immunodeficiency virus infection of cell arrested in the cell cycle. EMBO J 11:3053–3058

Miyoshi H et al (1998) Development of a self-inactivating lentivirus vector. J Virol 72:8150–8157

Miyoshi H et al (1999) Transduction of human CD34[+] cells that mediate long-term engraftment of NOD/SCID mice by HIV vectors. Science 283:682–686

Miyoshi H, Takahashi M, Gage FH, Verma IM (1997) Stable and efficient gene transfer into the retina using an HIV-based lentiviral vector. Proc Natl Acad Sci USA 94:10319–10323

Naldini L et al (1996) In vivo gene delivery and stable transduction of nondividing cells by a lentiviral vector. Science 272:263–267

Olsen JC (1998) Gene transfer vectors derived from equine infectious anemia virus. Gene Ther 5:1481–1487

Poeschla EF, Wong-Staal F, Looney DJ (1998) Efficient transduction of nondividing human cells by feline immunodeficiency virus lentiviral vectors. Nature Med 4:354–357

Roe T, Reynolds TC, Yu G, Brown PO (1993) Integration of murine leukemia virus DNA depends on mitosis. EMBO J 12:2099–2108

Sutton R et al (1998) Human immunodeficiency virus type 1 vectors efficiently transduce human hematopoietic stem cells. J Virol 72: 5781–5788

Sutton R, Uchida N, Brown P (1999) Transduction of human progenitor hematopoietic stem cells by human immunodeficiency virus type 1-based vectors is cell cycle dependent. J Virol 73:3649–3658

Takahashi M, Miyoshi H, Verme IM, Gage FH (1999) Rescue from photoreceptor degeneration in the rd mouse by human immunodeficiency virus vector-mediated gene transfer. J Virol 73:7812–7816

Uchida N et al (1998) HIV, but not murine leukemia virus, vectors mediate high efficiency gene transfer into freshly isolated G0/G1 human hematopoietic stem cells. Proc Natl Acad Sci USA 95:11939–11944

von Schwedler U, Kornbluth RS, Trono D (1994) The nuclear localization signal of the matrix protein of human immunodeficiency virus type 1 allows the establishment of infection in macrophages and quiescent T-lymphocytes. Proc Natl Acad Sci USA 91:6992–6996

White SM et al (1999) Lentivirus vectors using human and simian immunodeficiency virus elements. J Virol 73:2832–2940

Woods NB et al. (2000) Lentiviral gene transfer into primary and secondary NOD/SCID repopulating cells. Blood, in press

Zufferey R et al (1997) Multiply attenuated lentiviral vector achieves efficient gene delivery in vivo. Nat Biotechnol 5:871–875

Zufferey R et al (1998) Self-inactivating lentivirus vector for safe and efficient in vivo gene delivery. J Virol 72:9873–9880

3 Origin and Developmental Plasticity of Haematopoietic Stem Cells

A.M. Müller, H. Geiger, G. Eichholz, F. Harder

3.1 Introduction

Cellular homeostasis of various solid tissues as well as those composed of single cells such as skin, intestine and the haematopoietic system is maintained by a hierarchical structured cell system, with stem cells as their cellular origin. Stem cells are functionally defined by their capacity to self-renew and to lead to multi-lineage differentiation; hence, stem cells have the ability to generate both new stem cells and to clonally regenerate all the different cell types that constitute a stem cell system.

Stem cell systems consist in principle of three different cell types: (1) stem cells capable of unlimited self-renewal and competence to proliferate and to differentiate into multiple cell lineages; (2) progenitor cells with no self-renewal potential but able to proliferate and oligo-lineage differentiate; and (3) terminally differentiated cells with no inherent self-renewal ability and limited potential to differentiate.

Stem cells are the only permanent cell types, and under normal conditions progenitor cell clones and differentiated mature cells die out to be replaced by new progenitor cell clones differentiating from the stem cell pool. The highest burden of cell divisions is placed on the progenitor cells, which have no self-renewal potential and therefore have only a limited life span. The genome-wide number of mutations in a cell is coupled to the number of cell divisions it has completed (Turker 1998). Thus, the likelihood of a deleterious mutation occurring in progenitor cells is greater than in stem cells. If a mutation occurs in a progenitor clone, this mutated clone will finally cease to exist, since all but the stem cells are transient cell populations. As a consequence, cell systems which are generated and maintained by stem cells are less prone to the accumulation of mutations in comparison with other cell systems where mature and differentiated cells are able to proliferate. Further biological reasons for the existence of stem cell systems lie in the observation that the proliferation of progenitor cells and the function which is a feature of mature cells are separated. This opens up room for further specialisation of differentiated cells as seen in cells specialised beyond a point where they are able to divide, such as mammalian erythrocytes, thrombocytes, and keratinised epithelial cells of the skin.

Indispensable to the successful clinical application of stem cells as a source for tissue repair and replacement is an understanding of the properties and basic nature of stem cells, and how self-renewal and differentiation are regulated. The haematopoietic system is by far the best-known stem cell system, and much progress has been made in analysing the origin, cell surface characteristics and biological activity of haematopoietic stem cells (HSC) throughout development. In utero transplantations of HSC have been carried out in animal systems to successfully correct defects in the stem cell compartment of the unborn foetus, demonstrating the power of this transfer strategy (Fleischman et al. 1982; Toles et al. 1989). This work laid the foundation for in utero transplantation of human HSC into human foetuses to correct inherited disorders (for review, see Flake and Zanjani 1999). One strategy to get further insights into the biology of stem cells is to analyse events early in development, when stem cells originate and migrate to sites where they are found in the adult. In order to further define the prerequisites for successful heterochronic transplantation strategies (i.e. transplantation of adult-type stem cells into foetuses) utilising HSC, this chapter will

discuss recent work analysing the development of the vertebrate haematopoietic system and address the question of how much developmental potential resides in HSC.

3.2 Development of the Haematopoietic System

Basic theories describing phenomena in nature moderate from complex
to more straightforward as our knowledge increases. The same is true as
a description of the developing haematopoietic system. Historically, due
to its easy accessibility the chick embryo was the animal model in which
fundamental principles describing the origin of the haematopoietic system were established. The yolk sac (YS) was recognised as a haematopoietic organ in avian embryos in the early years of this century
(Dantschakoff 1908). It was later shown that the first haematopoietic
cells generated in the developing embryo originate from the extra-embryonic mesoderm of the YS via bipotential haemangioblasts, which can
develop into haematopoietic and endothelial cell lineages (Sabin 1920;
Choi et al. 1998; Eichmann et al. 1998). Accordingly, YS-derived stem
cells were regarded to be the only source of both the primitive (embryonic-type) and the definitive (foetal- and adult-type) HSC and haematopoietic systems (Moore and Owen 1965, 1967). But this "classical
theory" of developmental haematopoiesis was soon to be contested.
Elegant studies aimed at defining the origin of the definitive haematopoietic system in the chick embryo, which utilised the grafting of
quail embryos onto chick YS of comparable developmental stages, led
to a different conclusion. The haematopoietic cells in foetal and adult
spleen and thymus of quails developing from quail embryos grafted
onto chick YS were found to be derived from the quail and not to have
been colonised by the chick YS (Dieterlen-Lievre 1975). These and
many other experiments showed that there are two waves of haematopoiesis occurring in birds; one transient wave giving rise to primi-
tive-type haematopoiesis in the YS, and a second wave, originating de
novo from an intra-embryonic source which generates the definitive
haematopoietic system in birds. Further experiments identified the wall
of the aorta as the emergence site of multipotent haematopoietic progenitors (Cormier et al. 1986; Cormier and Dieterlen-Lievre 1988). It
was discovered that the wall of the aorta with surrounding intra-embry-

onic mesenchyme was very rich in haematopoietic progenitors, the first intra-embryonic site with haematopoietic activity, and hence regarded as the origin of the definitive haematopoietic system.

Comparison of the haematopoietic development in different species revealed a high degree of conservation in the basic processes occurring during early haematopoietic development between the species. As with birds, two waves of haematopoiesis exist in amphibians. In the *Xenopus* embryo, the ventral blood islands (VBIs) and the dorsal lateral plate (DLP) are active in haematopoiesis. The VBIs are located ventrally and produce both embryonic erythrocytes and thymocytes in early larvae, whereas those in late larvae and adults mainly come from the DLP located adjacent to the pronephros and the pronephric ducts (Kau and Turpen 1983; Maeno et al. 1985; Turpen et al. 1997; Huber and Zon 1998).

As mentioned above, strong evidence is accumulating for the existence of two distinct haematopoietic stem cell populations in amphibians and birds. These two populations are active during development, but are separated in space and time. This concept guided the analysis of the developing murine and human haematopoietic systems. In line with other vertebrate species, the murine and human YS is the first tissue endowed with haematopoietic activity during development, and it would therefore be logical to assume that the YS is the origin of HSC active throughout ontogeny. Haematopoietic cells capable of reconstituting recipient animals have been detected in the murine YS from embryonic day 8 (E8) onwards (Weissman et al. 1977; Toles et al. 1989; Yoder et al. 1997a,b) and in the YS and caudal half of the E9 embryo (Tyan and Herzenberg 1968). But reanalyses of the tissue which first gives rise to haematopoietic cells capable of long-term repopulation of the entire haematopoietic system of adult recipients revealed a region surrounding the dorsal aorta as the first intra-embryonic site with HSC activity (Müller et al. 1994; Dzierzak et al. 1995; Medvinsky and Dzierzak 1996). This region was named the AGM region, comprises aorta (A), gonad (G) and mesonephros (M) and evolves from the para-aortic splanchnopleure (P-Sp). Studies analysing the haematopoietic activity in early embryos detected multipotent haematopoietic progenitors in the P-Sp but not in the YS, from E8.5 onwards (Godin et al. 1993, 1995). Colony forming units-spleen (CFU-s), a class of multi-lineage haematopoietic progenitors, were first detected at E9 in the AGM-region but not in foetal liver or other parts of the embryo (Medvinsky et al. 1993, 1996). Furthermore, using an in vitro organ culture system multipotent progenitors including lymphoid progenitors developed autonomously in P-Sp explants (Cumano et al. 1996). Additionally, HSC with the potentiality of long-term and multi-lineage repopulation, originate in the AGM-region but not in other tissues (Medvinsky and Dzierzak 1996). These observations strongly suggest that definitive HSC originate in murine embryos in the region surrounding the dorsal aorta.

Foetal hepatic haematopoiesis is initiated by haematopoietic cells migrating from extra- as well as intra-embryonic sites which home in on the developing liver beginning at E9 (Houssaint 1981). But HSC are not detected there until E11 (Müller et al. 1994). Between early day 9 and day 14 of gestation, haematopoiesis in the foetal liver is characterised by a massive expansion of haematopoietic progenitors and stem cells (Metcalf and Moore 1971). Later in development at E14, haematopoietic progenitor cells migrate to the spleen (Borghese 1959), and on the 16th day of gestation the bone marrow commences its haematopoietic function (de Aberle 1928).

In common with other vertebrate species, haematopoiesis starts in human embryos initially in the YS at the 3rd week of gestation and shifts to the liver at week 5 (Peault 1996). Intra-embryonic haematopoiesis develops in the AGM region where multipotent progenitor cells first appear. They are associated with the ventral endothelium of the aorta in 5-week old embryos and can be detected in the bone marrow at gestational week 10 (Charbord et al. 1996; Tavian et al. 1996).

Together, these experiments strongly suggest that despite the initial appearance of haematopoiesis in the YS, HSC giving rise to the definitive haematopoietic system originate in the AGM region from where they migrate to foetal liver, spleen and bone marrow.

3.3 Developmental Changes in the Haematopoietic System

The major task for the haematopoietic system early in development is the supply and the removal of respiratory gases (oxygen and carbon dioxide) and accordingly, the first cells of haematopoietic origin generated at E7.5 of murine development are erythrocytes (Fig. 1). Macrophages, which play a role in both embryonic tissue remodelling and

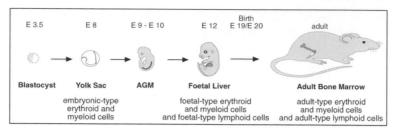

Fig. 1. Development of the murine haematopoietic system. The development of the murine embryo is shown starting from the blastocyst, depicting different tissues and organs (yolk sac, AGM region, foetal liver and bone marrow) active in murine haematopoiesis and their developmental stage-specific haematopoietic activities

inflammatory responses, first appear in E10 embryos in the YS, liver and urogenital ridges (Cline and Moore 1972; de Felici et al. 1986; Morris et al. 1991), whereas granulocytes and megakaryocytes appear at about E11 in the foetal liver (Metcalf and Moore 1971). The first cells of lymphoid origin are T cells seeding the foetal thymus on the 11th day of gestation (Auerbach 1961; Owen and Ritter 1969) and B cells are first detected in the foetal liver on day 12 of gestation and expand up to E16 (Chang et al. 1992).

Different developmental stages produce different haematopoietic cell subsets (for review, see Bonifer et al. 1998). Ontogeny-related changes reflect most likely the different duties haematopoietic cells have to fulfil during development. Mammalian embryonic-type erythrocytes generated during YS haematopoiesis are, in contrast to their foetal and adult counterparts, nucleated and express the embryonic-type globin genes which have a high oxygen binding activity (Craig and Russell 1964; Russell 1979). In mammals, the presence of a nucleus in embryonic erythrocytes might reflect an evolutionary link to lower vertebrates. Ontogeny-related differences are seen in the distinct macrophage populations developing during ontogeny, which differ immunophenotypically, in ultrastructure and in their gene expression pattern (Naito et al. 1990; Morioka et al. 1994; Faust et al. 1997). A similar heterogeneity is detected at the level of antigen receptors expressed on foetal and adult-type T and B cells. During foetal development, T cells bearing γδ receptors appear before αβ-positive T cells can be detected. Foetal thymus-derived T cells utilise an alternative set of V (variable) genes to their adult counterparts (Garman et al. 1986; Elliott et al. 1988) and were found to carry no N (inserted) nucleotides at the V-(D)-J junctions (Lafaille et al. 1989). Developmental stage-specific differences have been reported in the B cell lineage where CD5$^+$ (B1a) B cells are predominantly produced during foetal liver haematopoiesis and are believed to belong to a type of the B cell lineage different from conventional B cells (Hayakawa et al. 1985; Herzenberg et al. 1992).

Functional and phenotypic differences have not only been seen at the level of mature cells but also in the stem and progenitor cell compartment. Stem cells from foetal and adult tissues show differences in cell-cycle characteristics, turn-over time and cell surface attributes (Lansdorp et al. 1993; Lansdorp 1995; Morrison et al. 1995; Auerbach et al. 1996; Rebel et al. 1996). For example, HSC from the foetal liver and the adult bone marrow are recognised and can be highly enriched with the SCA-1, and anti-ckit monoclonal antibodies and are negative for lineage (LIN) markers, expressed on committed cells. In contrast to adult bone marrow-derived HSC, which are LIN$^-$, SCA-1$^+$, ckit$^+$ and are not recognised by the monoclonal antibodies AA4.1 and anti-Mac-1, foetal-derived HSC stain positive with the AA4.1 and the anti-Mac-1 antibodies (Rebel et al. 1996). Ontogeny related changes are further demonstrated by a switch in the thymic lymphocyte maturation potential of HSC. The potential to develop into Vγ3$^+$ T cells was tested in foetal liver- and bone marrow-derived HSC, and it was established that only foetal but not adult bone marrow-derived HSC have the capacity to differentiate into Vγ3$^+$ T cells in the foetal thymic microenvironment (Ikuta et al. 1990). Further examples are the favourable repopulation kinetics of foetal HSC over adult HSC as analysed by competitive repopulation analysis (Harrison et al. 1997) and the difference in homing patterns between foetal- and adult-type HSC (Zanjani et al. 1993).

In general, the more we know about foetal- and adult-type HSC, the more differences are recognised between these two cell populations; however, the molecular mechanisms regulating these changing developmental-specific properties are presently unknown.

3.4 Plasticity in the Developmental Potential of Haematopoietic Stem Cells

As outlined in the previous section, different developmental stages produce characteristic subsets of stage-specific haematopoietic cells which differ in many aspects. This observation raises several questions: How is the differentiation potential of HSC regulated and what are the contributions of intrinsic and extrinsic mechanisms? Furthermore it is of particular importance for the clinical applications to analyse the developmental potential and the migration and homing characteristics of in utero injected haematopoietic stem and progenitor cells. To address these issues, we exposed adult-type stem cells to embryonic microenvironments by injecting them into early murine embryos (Geiger et al. 1998). The earliest developing embryonic structure which can be injected is the preimplantation blastocyst. Since the lumen of the blastocyst provides space for only 20–40 cells, the HSC which are present in the LIN⁻, ckit⁺ and SCA-1⁺ cell population of the bone marrow had to be highly enriched (Okada et al. 1992). By using limiting dilution analyses, we established that 40 LIN⁻, ckit⁺ and SCA-1⁺ cells are sufficient to multi-lineage repopulate the complete haematopoietic system of recipient animals in the long term, thus demonstrating that HSC are present in this cell subset.

Analysing the donor contribution in female embryos developing from blastocysts injected with 20–40 HSC of male animals revealed the presence of male donor cells at E11.5 and E15.5 in YS and foetal liver (Fig. 2A; embryos C–F). Detection of donor-derived cells in the bone marrow, lymph nodes and peripheral blood of an 8-week-old animal demonstrated that donor-derived haematopoietic cells survive the entire ontogeny (Fig. 2B).

The results shown so far reveal the presence of donor cells in developing murine embryos and in an adult animal, but they do not cast any light on the nature of the donor cells. To find out whether the donor cells found in chimeric embryos have proliferation and differentiation potential, donor-specific progenitor cell assays were performed. HSC isolated from Neo- and LacZ-double transgenic ROSA26 animals (Friedrich and Soriano 1991) were injected into blastocysts conceived by non-transgenic animals. Foetal livers of developing chimeric animals were isolated and single cell suspensions were seeded in methylcellulose for

A

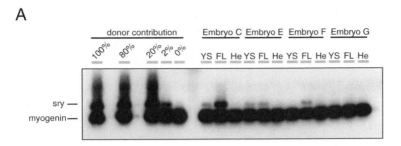

B

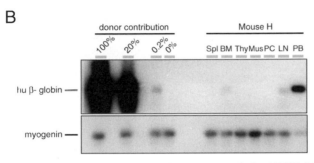

Fig. 2A,B. Donor contribution of bone marrow-derived HSC following injection into blastocysts. HSC from male donor animals or human β-globin transgenic animals were purified and injected into blastocysts as described (Geiger et al. 1998). Recipient female animals were analysed on E11.5 (*embryo C*) and on E15.5 (*embryos E–G*) for male donor contribution by sry-/myogenin-specific PCR (**A**) (Müller and Dzierzak 1993) or by human β-globin-/myogenin-specific PCR (Geiger et al. 1998) at 8 weeks of age (**B**). Relative donor contribution was determined by diluting male and human β-globin transgenic genomic DNA into female and non-transgenic genomic DNA. Dilutions are as indicated. Autoradiograms of Southern blot analyses are shown. *YS*, yolk sac; *FL*, foetal liver; *He*, head; *Spl*, spleen; *BM*, bone marrow; *Thy*, thymus; *Mus*, muscle; *PC*, peritoneal cells; *LN*, lymph nodes; *PB*, peripheral blood

A

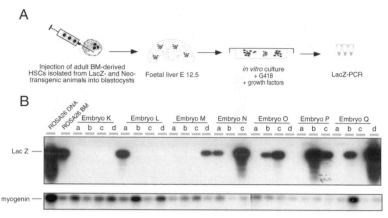

Fig. 3A,B. Donor-derived haematopoietic progenitor cells in chimeric foetal livers. HSC isolated from ROSA26 transgenic animals carrying a lacZ- and a neomycin-transgene (Friedrich and Soriano 1991) were injected into blastocysts developing from mating of wild-type animals (**A**). On E12.5, foetal livers were isolated, and single cell suspensions were seeded into methylcellulose cultures supplemented with haematopoietic growth factors and G418 for 16 days. Individual colonies (*a–d*) grown from foetal livers of a non-injected embryo (*K*) and injected embryos (*L–Q*) were isolated, genomic DNA was isolated, and lacZ/myogenin donor-specific PCR was performed (**B**)

16 days in the presence of G418 and haematopoietic growth factors (Fig. 3A). Due to the G418 selection, cells without the neomycin resistance gene eventually die. However only small-sized colonies were growing from chimeric foetal livers in our methylcellulose cultures. To determine their cellular origin, individual colonies were isolated and donor-specific lacZ-polymerase chain reaction (PCR) was performed on genomic DNA (Fig. 3B). PCR analysis demonstrated that G418-resistant and lacZ-transgenic colonies are present, indicating that HSC following the injection into blastocysts migrate to the developing foetal liver and generate haematopoietic progenitor cells capable of clonal proliferation in methylcellulose cultures.

In order to determine whether adult-type HSC isolated from the bone marrow retained their adult identity in the developing murine embryos, HSC from adult human β-globin transgenic animals were injected into blastocysts. The human β-globin transgene locus shows tissue- and

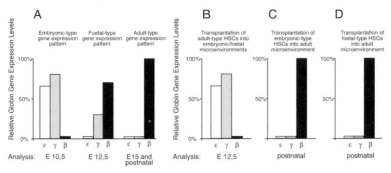

Fig. 4A–D. Developmental-specific gene expression pattern following heterochronic transplantations. The relative gene expression pattern of human β-globin genes is shown in transgenic animals during development. **A** Composite representation of various studies analysing the gene expression of the human ε-, γ- and β-globin transgenes during development (Strouboulis et al. 1992; Gribnau et al. 1998). The maximum expression levels of individual human ε-, γ- and β-globin transgenes were set to 100%. **B–D** show the analysis of the gene expression pattern of human β-globin transgenes in progeny of haematopoietic progenitors and stem cells following heterochronic transplantations. Adult-type HSC of human globin transgenic animals were isolated and injected into blastocysts, and gene expression of individual members of the human β-globin locus was analysed in the developing chimeric embryos at E12.5 in the foetal liver by reverse transcription (RT) PCR (**B**). YS-derived (**C**) and AGM- and foetal liver-derived haematopoietic progenitors (**D**) of human β-globin transgenic animals were injected intravenously into irradiated animals; spleen colonies were isolated on days 8 and 11 post-transplantation, and human globin-specific RT PCR was done on RNA isolated from individual colonies. Relative expression levels of ε-, γ- and β-human globin transgenes are shown

developmental-specific gene expression in mice (Strouboulis et al. 1992). The human ε-globin gene is expressed during embryonic development, whereas the human γ-globin gene behaves as an embryonic-, and foetal-expressed gene, and the human β-globin gene is active during foetal and adult developmental life of the animals (Fig. 4A). Thus, gene expression analysis of members of the human globin locus allows us to distinguish between embryonic-, foetal- and adult-type gene expression programs in heterochronically injected HSC. HSC from human β-globin transgenic animals were isolated and injected into blastocysts, and the gene expression pattern of the transgene locus was consequently assessed in chimeric foetal livers by human globin-specific reverse transcription (RT) PCR. Rather surprisingly, we detected transcripts of the embryonic and foetal expressed human ε-, and γ-globin genes, but not the adult- and foetal-type human β-gene (Fig. 4B). Conversely, if YS-, AGM- and foetal liver-derived progenitors and HSC were injected into adult mice, they generated splenic colonies which express the adult-type human β-globin transgene, not the embryonic- or foetal-type globin genes (Fig. 4C,D). The results from these experiments suggest that it is the microenvironment that dictates which gene expression program becomes activated in haematopoietic progenitors and stem cells. Furthermore, interactions between the microenvironment and HSC determine the developmental fate of the transplanted stem cells.

3.5 Summary and Perspectives

Here we report that HSC isolated from the bone marrow of adult mice can generate chimeric animals following injection into preimplantation blastocysts. Donor cells could be detected in haematopoietic organs and tissues in foetal and adult developing stages, implying the homing of the injected HSC to sites active in embryonic, foetal and adult haematopoiesis. The exact nature of the donor cells in chimeric tissues is presently not known. However, we could detect haematopoietic progenitor activity capable of differentiating in methylcellulose cultures, and we were able to demonstrate donor-derived globin gene expression in the foetal liver, suggesting the presence of progenitor cells and committed erythroid cells. Whether donor-derived HSC are present in chimeric embryos and whether the weak signal in bone marrow and lymph nodes of adult chimeric animals originates from lymphoid cells is presently not known.

Interestingly, analysing developmental stage-specific gene expression patterns in offspring of the injected donor cells reveals a broad plasticity within the HSC compartment. The injected adult-type HSC produce within the embryonic microenvironment cells which express the embryonic or foetal transgenes and silence the adult-type human globin transgene, whereas embryonic and foetal-type haematopoietic progenitor cells transcribe in the microenviroment provided by the adult spleen the adult-type human β-globin gene but not the embryonic ε- and embryonic/foetal γ-globin transgenes. This result indicates that developmental-specific changes in the gene expression pattern are regulated by microenvironments and are reversible. Thus, the microenvironment has greater regulatory influence than previously thought. It is not, however, clear whether the environment influences at the level of stem cells or on lineage restricted cells, nor is the molecular nature of the receptors recognising the developmental age of the microenvironment and are the signal transduction pathways understood.

This apparent plasticity should make it possible in the future to rejuvenate stem cells from adults to gain foetal characteristics. Foetal HSC show superior repopulation characteristics, possess wider differentiation potentials and lower incidence of graft versus host disease (GvHD) than their adult counterparts, and it will be of great importance to determine whether and how these properties can be (re)-activated in adult-type stem cells. Furthermore, the emerging developmental plasticity of stem cells as seen in the studies described here and in reports demonstrating haematopoietic potential in neuronal and muscle stem cells (Bjornson et al. 1999; Gussoni et al. 1999) has great theoretical and practical implications, since stem cells seem to posses a wider developmental potential than previously thought. In this respect, it is intriguing to think of stem cells as cells not restricted to one stem cell system, but of cells with a broader differentiation spectrum if provided with the right microenvironmental stimuli. This perspective opens the door for more and exciting basic research activities and widens the potential use of stem cells in cell and tissue replacement strategies. Further research is needed to uncover the full developmental potential of stem cells.

Acknowledgements. The authors thank Bruce Jordan for comments and critically reading the manuscript. This work was supported by the Max-Planck Gesellschaft, Germany, the DFG, Germany and the DFG-funded programme project grant SFB 465.

References

Auerbach R (1961) Experimental analysis of the origin of the cell types in the development of the mouse thymus. Dev Biol 3:336–364

Auerbach R, Huang H, Lu L (1996) Hematopoietic stem cells in the mouse embryonic yolk sac. Stem Cells 14:269–280

Bjornson CRR, Rietze RL, Reynolds BA, Magli MC, Vescovi AL (1999) Turning brain into blood: a hematopoietic fate adopted by adult neural stem cells in vivo. Science 283:534–537

Bonifer C, Faust N, Geiger H, Müller AM (1998) Developmental changes in the differentiation capacity of haematopoietic stem cells. Immunol Today 19:236–241

Borghese E (1959) The present state of research on WW mice. Acta Anat 36:185–220

Chang Y, Paige CJ, Wu GE (1992) Enumeration and characterization of DJH structures in mouse fetal liver. EMBO J 11:1891–1899

Charbord P, Tavian M, Humeau L, Peault B (1996) Early ontogeny of the human marrow from long bones: an immunohistochemical study of hematopoiesis and its microenvironment. Blood 87:4109–4119

Choi K, Kennedy M, Kazarov A, Papadimitriou JC, Keller G (1998) A common precursor for hematopoietic and endothelial cells. Development 125:725–732

Cline MJ, Moore MAS (1972) Embryonic origin of the mouse macrophage. Blood 39:842–849

Cormier F, Dieterlen-Lievre F (1988) The wall of the chick embryo aorta harbours M-CFC, G-CFC, GM-CFC and BFU-E. Development 102:279–285

Cormier F, de Paz P, Dieterlen-Lievre F (1986) In vitro detection of cells with monocytic potentiality in the wall of the chick embryo aorta. Dev Biol 118:167–175

Craig ML, Russell ES (1964) A developmental change in hemoglobins correlated with an embryonic red cell population in the mouse. Dev Biol 10:191–201

Cumano A, Dieterlen-Lievre F, Godin I (1996) Lymphoid potential, probed before circulation in mouse, is restricted to caudal intraembryonic splanchnopleura. Cell 86:907–916

Dantschakoff W (1908) Untersuchungen über die Entwicklung des Blutes und Bindegewebes bei den Vögeln. Anat Hefte 37:471–589

de Aberle SB (1928) A study of the hereditary anaemia of mice. Am J Anat 40:219–249

de Felici M, Heasman J, Wylie CC, McLaren A (1986) Macrophages in the urogential ridge of the mid-gestation mouse fetus. Cell Differ 18:119–129

Dieterlen-Lievre F (1975) On the origin of haemopoietic stem cells in the avian embryo: an experimental approach. J Embryol Exp Morphol 33:607–619

Dzierzak E, Müller A, Sinclair A, Miles C, Gillett N, Daly B, Sanchez M-J, Medvinsky A (1995) Hematopoietic stem cell development in the mouse embryo. In: Proceedings of the 9th conference on hemoglobin switching, pp 109–121

Eichmann A, Corbel C, Le Douarin NM (1998) Segregation of the embryonic vascular and hemopoietic systems. Biochem Cell Biol 76:939–946

Elliott JF, Rock EP, Patten PA, Davis MM, Chien Y-h (1988) The adult T-cell receptor delta-chain is diverse and distinct from that of fetal thymocytes. Nature 331:627–631

Faust N, Huber MC, Sippel AE, Bonifer C (1997) Different macrophage populations develop from embryonic/fetal and adult hematopoietic tissues. Exp Hematol 25:432–444

Flake AW, Zanjani ED (1999) In utero hematopoietic stem cell transplantation: ontogenic opportunities and biologic barriers. Blood 94:2179–2191

Fleischman RA, Custer RP, Mintz B (1982) Totipotent hematopoietic stem cells: normal self-renewal and differentiation after transplantation between mouse fetuses. Cell 30:351–359

Friedrich G, Soriano P (1991) Promoter traps in embryonic stem cells: a genetic screen to identify and mutate developmental genes in mice. Genes Dev 5:1513–1523

Garman RD, Doherty PJ, Raulet DH (1986) Diversity, rearrangement, and expression of murine T cell gamma genes. Cell 45:733–742

Geiger H, Sick S, Bonifer C, Müller AM (1998) Globin gene expression is reprogrammed in chimeras generated by injecting adult hematopoietic stem cells into mouse blastocysts. Cell 93:1055–1065

Godin IE, Garcia-Porrero JA, Coutinho A, Dieterlen-Lievre F, Marcos MAR (1993) Para-aortic splanchnopleura from early mouse embryos contains B1a cell progenitors. Nature 364:67–70

Godin I, Dieterlen-Lievre F, Cumano A (1995) Emergence of multipotent hemopoietic cells in the yolk sac and paraaortic splanchnopleura in mouse embryos, beginning at 8.5 days postcoitus. Proc Natl Acad Sci USA 92:773–777

Gribnau J, de Boer E, Trimborn T, Wijgerde M, Milot E, Grosveld F, Fraser P (1998) Chromatin interaction mechanism of transcriptional control in vivo. EMBO J 17:6020–6027

Gussoni E, Soneoka Y, Strickland CD, Buzney EA, Khan MK, Flint AF, Kunkel LM, Mulligan RC (1999) Dystrophin expression in the mdx mouse restored by stem cells transplantation. Nature 401:390–394

Harrison DE, Zhong RK, Jordan CT, Lemischka IR, Astle CM (1997) Relative to adult marrow, fetal liver repopulates nearly five times more effectively long-term than short-term. Exp Hematol 25:293–297

Hayakawa K, Hardy RR, Herzenberg LA, Herzenberg LA (1985) Progenitors for Ly-1 B cells are distinct from progenitors for other B cells. J Exp Med 161:1554–1568

Herzenberg LA, Kantor AB, Herzenberg LA (1992) Layered evolution in the immune system. A model for the ontogeny and development of multiple lymphocyte lineages. Ann NY Acad Sci 651:1–9

Houssaint E (1981) Differentiation of the mouse hepatic primordium. II. Extrinsic origin of the haemopoietic cell line. Cell Differ 10:243–252

Huber TL, Zon LI (1998) Transcriptional regulation of blood formation during Xenopus development. Semin Immunol 10:103–109

Ikuta K, Kina T, MacNeil I, Uchida N, Peault B, Chien Y-h, Weissman IL (1990) A developmental switch in thymic lymphocyte maturation potential occurs at the level of hematopoietic stem cells. Cell 62:863–874

Kau CL, Turpen JB (1983) Dual contribution of embryonic ventral blood island and dorsal lateral plate mesoderm during ontogeny of hemopoietic cells in Xenopus laevis. J Immunol 131:2262–2266

Lafaille JJ, DeCloux A, Bonneville M, Takagaki Y, Tonegawa S (1989) Junctional sequences of T cell receptor gamma delta genes: implications for gamma delta T cell lineages and for a novel intermediate of V-(D)-J joining. Cell 59:859–870

Lansdorp PM (1995) Developmental changes in the function of hematopoietic stem cells. Exp Hematol 23:187–191

Lansdorp PM, Dragowska W, Mayani H (1993) Ontogeny-related changes in proliferative potential of human hematopoietic cells. J Exp Med 178:787–791

Maeno M, Tochinai S, Katagiri C (1985) Differential participation of ventral and dorsolateral mesoderms in the hemopoiesis of Xenopus, as revealed in diploid-triploid or interspecific chimeras. Dev Biol 110:503–508

Medvinsky A, Dzierzak E (1996) Definitive hematopoiesis is autonomously initiated by the AGM region. Cell 86:897–906

Medvinsky AL, Samoylina NL, Müller AM, Dzierzak EA (1993) An early pre-liver intraembryonic source of CFU-S in the developing mouse. Nature 364:64–67

Medvinsky AL, Gan OI, Semenova ML, Samoylina NL (1996) Development of day-8 colony-forming unit-spleen hematopoietic progenitors during early murine embryogenesis: spatial and temporal mapping. Blood 87:557–566

Metcalf D, Moore MAS (1971) Embryonic aspects of haematopoiesis. In: Neuberger A, Tatum EL (eds) Haematopoietic cells. North-Holland, Amsterdam, pp 172–271

Moore MAS, Owen JJT (1965) Chromosome marker studies on the development of the haematopoietic system in the chicken embryo. Nature 208:956–990

Moore MAS, Owen JJT (1967) Stem-cell migration in developing myeloid and lymphoid systems. Lancet ii:658–659

Morioka Y, Naito M, Sato T, Takahashi K (1994) Immunophenotypic and ultrastructural heterogeneity of macrophage differentiation in bone marrow and fetal hematopoiesis of mouse in vitro and in vivo. J Leukoc Biol 55:642–651

Morris L, Graham CF, Gordon S (1991) Macrophages in haemopoietic and other tissues of the developing mouse detected by the monoclonal antibody F4/80. Development 112:517–526

Morrison SJ, Hemmati HD, Wandycz AM, Weissman IL (1995) The purification and characterization of fetal liver hematopoietic stem cells. Proc Natl Acad Sci USA 92:10302–10306

Müller AM, Dzierzak EA (1993) ES cells have only a limited lymphopoietic potential after adoptive transfer into mouse recipients. Development 118:1343–1351

Müller AM, Medvinsky A, Strouboulis J, Grosveld F, Dzierzak E (1994) Development of hematopoietic stem cell activity in the mouse embryo. Immunity 1:291–301

Naito M, Takahashi K, Nishikawa S (1990) Development, differentiation, and maturation of macrophages in the fetal mouse liver. J Leukoc Biol 48:27–37

Okada S, Nakauchi H, Nagayoshi K, Nishikawa S-I, Miura Y, Suda T (1992) In Vivo and In Vitro Stem Cell Function of *c-kit-* and Sca-1-Positive Murine Hematopoietic Cells. Blood 80:3044–3050

Owen JJ, Ritter MA (1969) Tissue interaction in the development of thymus lymphocytes. J Exp Med 129:431–442

Peault B (1996) Hematopoietic stem cell emergence in embryonic life: developmental hematology revisited. J Hematother 5:369–378

Rebel VI, Miller CL, Thornbury GR, Dragowska WH, Eaves CJ, Lansdorp PM (1996) A comparison of long-term repopulating hematopoietic stem cells in fetal liver and adult bone marrow from the mouse. Exp Hematol 24:638–648

Russell ES (1979) Hereditary anemias of the mouse: a review for geneticists. Adv Genet 20:357–459

Sabin FR (1920) Studies on the origin of blood-vessels and of red blood-corpuscles as seen in the living blastoderm of chicks during the second day of incubation. Carnegie Inst Wash Publ (Contribut Embryol 9) 272:213–262

Strouboulis J, Dillon N, Grosveld F (1992) Developmental regulation of a complete 70-kb human beta-globin locus in transgenic mice. Genes Dev 6:1857–1864

Tavian M, Coulombel L, Luton D, Clemente HS, Dieterlen-Lievre F, Peault B (1996) Aorta-associated CD34$^+$ hematopoietic cells in the early human embryo. Blood 87:67–72

Toles JF, Chui DHK, Belbeck LW, Starr E, Barker JE (1989) Hemopoietic stem cells in murine embryonic yolk sac and peripheral blood. Proc Natl Acad Sci USA 86:7456–7459

Turker MS (1998) Estimation of mutation frequencies in normal mammalian cells and the development of cancer. Semin Cancer Biol 8:407–419

Turpen JB, Kelley CM, Mead PE, Zon LI (1997) Bipotential primitive-definitive hematopoietic progenitors in the vertebrate embryo. Immunity 7:325–334

Tyan ML, Herzenberg LA (1968) Studies on the ontogeny of the mouse immune system. II. Immunoglobulin-producing cells. J Immunol 101:446–450

Weissman I, Papaioannou V, Gardner R (1977) Fetal hematopoietic origins of the adult hematolymphoid system (Clarkson B ed). Cold Spring Harbor Laboratory Press, Cold Spring Harbor, pp 33–47

Yoder MC, Hiatt K, Dutt P, Mukherjee P, Bodine DM, Orlic D (1997a) Characterization of definitive lymphohematopoietic stem cells in the day 9 murine yolk sac. Immunity 7:335–344

Yoder MC, Hiatt K, Mukherjee P (1997b) In vivo repopulating hematopoietic stem cells are present in the murine yolk sac at day 9.0 postcoitus. Proc Natl Acad Sci USA 94:6776–6780

Zanjani ED, Ascensao JL, Tavassoli M (1993) Liver-derived fetal hematopoietic stem cells selectively and preferentially home to the fetal bone marrow. Blood 81:399–404

4 The New York Blood Center's Placental/Umbilical Cord Blood Program. Experience with a 'New' Source of Hematopoietic Stem Cells for Transplantation

P. Rubinstein, C.E. Stevens

4.1 Introduction

The first attempt to replace the hematopoietic system of a human with neonatal hematopoietic stem cells (HSC) from placental and umbilical cord blood (PCB) was made in 1988 by Boyse, Gluckman and their colleagues. Gluckman's transplant (Gluckman et al. 1989) was the culmination of much work during the two preceding decades that showed the presence of colony-forming progenitor cells and others capable of at least some self-replication in human PCB (Knudtzon 1974; Prindull et al. 1978; Fauser and Messner 1978; Nakahata and Ogawa 1982; Koike 1983; Besalduch 1985; Broxmeyer et al. 1989, 1992). Obvious logistic and immunogenetic considerations initially restricted the clinical application of PCB to patients who had newborn sibling donors. Over a dozen such transplants were conducted in a few years (Broxmeyer et al.

1991; Vilmer et al. 1992; Wagner et al. 1992a; Issaragrisil 1994; Kernan et al. 1994) from both human leukocyte antigen (HLA)-identical and non-identical sibling donors to patients with several different diseases, including inherited conditions and leukemia (Wagner et al. 1995; Gluckman et al. 1997).

Since the first applications were restricted to patients whose mothers could become pregnant again, the vast majority of patients in need of hematopoietic replacements would remain outside the possible benefit of this neonatal source of stem cells unless unrelated donors could be used. We argued in favor of the latter source on the grounds that most of the patients in need of marrow replacement are in this category and that the possible benefits to patients should outweigh the difficulties of establishing a "bank" of unrelated donor tissue for such patients (Rubinstein 1993; Rubinstein et al. 1993). The New York Blood Center's (NYBC) application for a relatively large level of funding from the NIH (National Institutes of Health) succeeded in 1992, after 3 years of discussions.

We had anticipated three types of potential advantages. The first focused on the "immunological immaturity" of neonatal donors, which might permit the use of partially incompatible donor tissue without unduly severe graft-versus-host disease (GvHD). This feature would help patients whose HLA types are uncommon in the population, particularly individuals from ethnic groups relatively underrepresented in the Bone Marrow Donor Registries. Second, we pointed out that much faster response times could become possible when using already studied, frozen grafts, in comparison with the donation by volunteers in Bone Marrow Registries, combined with the total absence of clinical risks to these donors. Thirdly, we discussed the much lower prevalence of some infections [particularly cytomegalovirus (CMV) and Epstein-Barr virus (EBV)] in neonates than in adult donors.

The demonstration of the usefulness of this source of transplantable HSC in unrelated patients (Kurtzberg et al. 1994; Rubinstein et al. 1994) was greeted with interest in the United States and other countries and led to the organization of additional programs that now collect, process and distribute PCB for unrelated (and under special conditions, also for related) patients. More recently, for-profit companies have begun offering private "speculative" storage of PCB for autologous or related-donor transplantation. These companies may provide scant guidance to their clients on the limited practical usefulness of such storage and on the potential problems of their use in autologous transplantation in childhood leukemia (Mahmoud et al. 1995; Ford et al. 1997; Gale et al. 1997; Rowley 1998; American Academy of Pediatrics Work Group on Cord Blood Banking 1999). These issues and the many technical, financial, and ethical issues affecting the functions of all PCB programs suggest the need for scientific and technical study at the regulatory agency level. The United States Food and Drug Administration (FDA) is currently engaged in such study.

The NYBC's Placental Blood Program was conceived from the start as a full service program, namely one that includes all aspects of the collection, processing, and clinical utilization of PCB donations. The program thus performs:

– Collection of PCB, maternal informed consent, clinical data, and pertinent family history
– Processing, cryopreservation and long-term storage of PCB units and test specimens
– Infectious and other disease testing as appropriate
– Immunogenetic typing
– Handling search requests from transplant centers on behalf of individual patients
– Identifying potentially useful donated PCB units for individual recipients
– Confirming HLA matching at the desired immunogenetic level
– Releasing the grafts to the transplant centers
– Maintaining follow-up information on the long-term performance of the PCB grafts and their recipients

Although we have reported on specific details of these various functions of the program (Rubinstein 1993; Rubinstein et al. 1993), the issues concerning informed consent and those related to the identification and preservation of units that may be stored for many years before transplantation will be specifically mentioned in this report.

4.1.1 Informed Consent Procedures

Two principal methods of collecting PCB are in current use, which condition the need for obtaining maternal consent in specifically different ways: one takes place during the third stage of labor, the other after the delivery of the placenta. The intervention of an additional maneuver modifies the conduct of labor, one of the safest procedures in medicine. While the modification is in itself minimal and carries little risk, it constitutes a deviation during a moment of labor which calls for evaluation of the infant and mother and in which distractions of the obstetrical team should probably be avoided as much as possible. An additional aspect of in utero collections is the conflict between the collection of as large a volume of left-over placental and umbilical cord blood as possible and the desirability to avoid clamping the cord in ways that may predispose to anemia in the infant (Bertolini et al. 1995). These characteristics require that the informed consent procedure include information on the possible, though unlikely, risks of in utero collecting when this is the technique selected. Obviously, such informed consent can only be obtained prior to labor. Published information on in utero collection methods thus far does not describe these aspects of the informed consent procedure in detail (Broxmeyer et al. 1989; Wagner et al. 1992b; Bertolini et al. 1995).

Collection after placental delivery (Rubinstein 1993; Rubinstein et al. 1993) does not interfere with the conduct of labor and implies no risks to either mother or infant. In this case, the consent procedures may focus simply on the information needed by the mother for granting the different permissions requested, irrespective of the type of collection intended. These permissions are:

- To obtain, test, and store the PCB and specimens
- To collect test results and report them to the mother and/or her physicians
- To gather specific data from the mother and the hospital charts
- To use the PCB for any patient that may need it
- To have the PCB irreversibly de-linked (identifying numbers removed) and used for research and quality assurance purposes, if found to be unsuitable for clinical use

The mother is also advised that, although her data are strictly handled as confidential and private, it may be required for Public Health authorities to have access to her ID and possibly to contact her family under situations defined by the health regulations.

4.1.2 Sample and Data Identification: A Single ID Number

Reliable testing and matching procedures require confidence on the validity of the relevance of the specimen or document to a specific donor or recipient. Although there is considerable experience, ID errors may occur in procedures, particularly in multi-step methods and those in which multiple aliquots must be obtained from primary sources. The NYBC methods involve pre-printed label sets, each designated to carry the same ID number on all labels within the set. These labels are used in the delivery suite laboratory where PCB is collected to identify the PCB unit after collection, the mother's sample, the data form, etc. The ID number is printed in both human- and machine-readable forms (normal text and bar-coded characters, respectively) and is used within the program by scanning the number for either data entry procedures or for copying to secondary (copy-cat) sample labels printed on demand. Because the scanners are site-specific within the program's computer network, they provide for real-time monitoring, verifying, and recording the progression of each sample within the system.

4.1.3 Cryopreservation and Instrumentation of Freezing

The traditional method of adding cryopreservative solutions of di-methyl-sulfoxide (DMSO) requires doubling the final volume of the units and reduces considerably the number of units that can be stored in a fixed reservoir space. Several methods for reducing the volume of the unit prior to addition of DMSO have been tried (Rubinstein et al. 1995; Denning-Kendall et al. 1996; Rebulla et al. 1998; Armitage et al. 1999; Tokushima et al. 1999;). We have reported data on engraftment of units whose volume was reduced by enhanced sedimentation with hydroxyethyl-starch (HES) (Rubinstein et al. 1995), and demonstrated no effects whatever of the probability of engraftment (Rubinstein et al.

1998). These results confirmed the in vitro data showing that the inevitable losses of mononuclear cells during the entire procedure are small (Rubinstein et al. 1995). The addition of 50%, rather than 20%, DMSO also helps reduce the final volume and increase the number of units that can be stored with this method.

The reduction to a constant final volume of 25 ml for every unit facilitates the task of optimizing storage. Thus, a plastic container was designed to fit this volume, which included providing a separable compartment allowing the user to access a part of the unit when cell-expansion procedures are desired, without compromising the rest of the frozen unit. Currently, this compartment holds 20% of the final unit volume. Because both the final volume and the concentration of DMSO are constant, the use of an optimized freezing rate curve becomes easier to program to automate the freezing of each unit separately using, for example, the robotic Thermogenesis machine. This allows shortening the length of the freezing cycle in comparison with standard controlled rate freezers, without compromising mononuclear cell viability or colony-forming cell function. It also avoids the need to freeze several units at one time in a single controlled rate freezing chamber and assumptions about the homogeneity of the freezing rates.

4.1.4 Program Milestones

The NYBC PCB Program supplied the first ever grafts for unrelated patients at Duke University in 1993. These were processed without volume reduction and cryoprotected using 20% DMSO, as were all grafts prepared through the winter of 1994, when the volume reduction protocols were completed and validated (Rubinstein et al. 1995). By the end of our NIH grant in 1995, the program had collected, processed, and stored 5000 units from donors at Mt. Sinai Hospital in New York and had provided units for 100 patients worldwide. In 1996, PCB collections were started at Brooklyn Hospital, where the ethnic stratification favored donors of African-American and Hispanic ancestry. Since 1996, the PCB program has operated under the IND (Investigational New Drug) program of the US FDA. Under IND regulations, the program was allowed to ask for reimbursement, which became its sole source of funding. The new freezing bags became available in 1998, validated and used in storing units for clinical use in thermogenesis freezers, beginning in 1999. By December, 1999, the program had shipped 1004 units from a then current inventory of 9796 to patients among 8804 submitting search requests. The units were shipped to 126 transplant centers in the United States, Europe, Latin America, Asia, and Oceania.

4.2 Clinical Endpoints

We have reported on the demographics and clinical diagnoses, as well as on the clinical endpoints of the first 562 transplants (Rubinstein et al. 1998). In this review, we will briefly address the results of all transplants currently reported to us with at least 100 days and up to 75 months of follow-up. The endpoint definition (summarized within the corresponding descriptions, below) and statistical methods used in this analysis were as given in Rubinstein et al. (1998). Briefly, engraftment, incidence of transplant-related events (TRE), relapse, and event-free survival are estimated with Kaplan-Meier's product-limit method (Kaplan and Meier 1958). We evaluate factors for their influence on these endpoints with univariate analyses using the log-rank statistic and the generalized Wilcoxon statistic (also known as the Breslow test), respectively. We also perform multivariate analyses with Cox logistic regression (Cox 1972; SPSS Advanced Statistical Software) under the assumption of proportional hazards with all analyzed covariates in the model (SPSS Advanced Statistical Software). Categorical data are compared using Fisher's exact test, Pearson's χ^2, or linear-by-linear χ^2 (all two-tailed) test (SPSS Advanced Statistical Software). All statistical routines are from the Statistical Program for the Social Sciences, SPSS Inc., Chicago, Ill.) (SPSS Advanced Statistical Software).

4.2.1 Engraftment

Myeloid engraftment is defined as the attainment of an absolute count of ≥ 500 neutrophils per microliter of blood for at least 3 consecutive days; it is measured on the 1st of those 3 days. Platelet engraftment is defined as reaching an absolute platelet count of $\geq 50,000/\mu l$, maintained for at least 7 days.

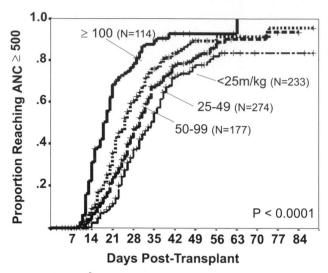

Fig. 1. Cell Dose ($\times 10^6$/kg) and time to ANC500

It has been clear for some time that the total nucleated cell dose (TNC) is a major determinant of the time to engraftment (Sullivan et al. 1989; Kurtzberg et al. 1996; Gluckman et al. 1997; Rubinstein et al. 1998; Visser et al. 1999), as also demonstrated here by Kaplan-Meier statistics (Fig. 1).

Interestingly, the three highest cell dosages all have similar Kaplan-Meier probability of reaching engraftment, and the difference between them is mostly a matter of speed. An interesting aspect of the cell dose in PCB transplantation has been identified recently (P. Wernet, personal communication): the presence of nucleated red blood cells (NRBC) in some PCB, occasionally at high concentrations relative to the white cell counts and thus able to contribute importantly to the TNC. This factor can cause unpleasant surprises when determining the post-thaw TNC of a PCB graft: since most nucleated erythrocytes lyse during thawing, the TNC can become very much lower. The problem is due to the inability of most current automated hematology analyzers to distinguish between NRBC and leukocytes. Fortunately, and despite their tendency to lysis, the pre-freezing dose of NRBC correlates with time to engraftment no less well than the dose of the other nucleated cells in the TNC (C.E.

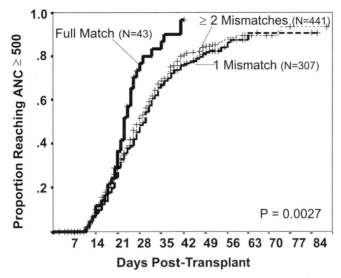

Fig. 2. HLA match and time to ANC500

Stevens, in preparation). Although the TNC dose is used traditionally to evaluate the hematopoietic content of a stem cell graft, our outcome data indicate that colony-forming progenitor cells are an even better surrogate (A.R. Migliaccio et al., in preparation). In a preliminary experiment, Visser et al. (1999) showed that a particular subset of PCB leukocytes bearing the CD34 marker correlates very closely with the colony-forming progenitor cells (Pearson's correlation coefficient, $r \geq 0.94$). Thus, the hematopoietic progenitor cell content may become easier and faster to evaluate, helping to improve clinical evaluation of the available dose in choosing between similar grafts for specific patients.

   The relationship between HLA match and PCB engraftment has been less easy to pin down. Our data, however, strongly indicate the presence of the expected association (Fig. 2, Table 1). Both univariate and multivariate analyses demonstrate that engraftment is possible with up to (at least) three major mismatches, but that the speed of engraftment is significantly lower in the HLA mismatched than in the "fully-matched" group. Platelet engraftment, on the other hand, is clearly associated with

Table 1. Uni- and multivariate analysis of relative risk for faster engraftment[a]

| Variable | Number of patients | Relative risk for time to ANC $\geq$500 (95% CI) | | | |
		Univariate	P	Multivariate	P
1. Diagnosed with CML, Fanconi anemia, or severe aplastic anemia					
No	644	1.8 (1.4–2.3)	0.001	1.5 (1.1–1.9)	0.004
Yes	134	1.0		1.0	
2. Age group:					
<2 years	157	2.3 (1.8–2.9)	0.001	0.97 (0.7–1.4)	0.9
2–11 years	358	1.3 (1.0–1.5)	0.02	0.84 (0.7–1.1)	0.15
$\geq$12 years	263	1.0		1.0	
3. TNCs per kilogram (pre-cryopreservation, $\times 10^6$):					
$\geq$100	113	3.2 (2.5–4.2)	0.001	3.2 (2.3–4.6)	0.001
50–99	175	1.8 (1.4–2.2)	0.001	1.8 (1.3–2.4)	0.001
25–49	267	1.3 (1.0–1.6)	0.02	1.3 (1.1–1.7)	0.017
	223	1.0		1.0	
4. Number of HLA-A, -B, -DR mismatches:					
None	43	1.7 (1.2–2.4)	0.001	1.8 (1.3–2.5)	0.001
$\geq$1	733	1.0		1.0	
US and non-US transplant centers:					
US	636	1.7 (1.3–2.1)	0.001	1.7 (1.4–2.2)	0.001
Non-US	142	1.0		1.0	

[a] The analysis excludes 67 patients whose time to ANC $\geq$500 is unknown (including 26 patients with SCID who were not evaluable because they had not undergone bone marrow ablation). The Cox regression also excludes 17 patients who died or relapsed before the first patient engrafted (day 9). Two patients who could not be categorized for HLA mismatch because of lack of high resolution DRB1 typing at transplantation were also excluded from the multivariate analysis.

Table 2. HLA and acute graft-vs-host disease in recipients of unrelated PCB transplants

Number of HLA mismatches[a]	Patients (n)	GvHD Grades 0–I (n)	Grade II (n)	Grades III–IV (n)	P value
None	36	26 (72%)	7 (19%)	3 (8%)	
1	227	128 (56%)	57 (25%)	42 (19%)	
≥2	323	154 (48%)	77 (24%)	92 (28%)	0.006
					(linear-by-linear association, P<0.001)
Total	586[b]	308 (52%)	141 (24%)	137 (24%)	

[a] Mismatches assigned based on serology with splits, for HLA-A and -B, and on DNA allele-resolution tests for DRB1.
[b] Table includes 550 patients who are known to have reached an ANC ≥500, 16 who had some evidence of engraftment but did not reach ANC ≥500, 15 with no evidence of engraftment, and 6 with no report to date on ANC status. One patient not typed for DRB1 at the allele level could not be categorized.

cell dose, age, acute GvHD, and incident infections, but the association with HLA match grade does not reach significance as yet.

4.2.1.1 Graft-Versus-Host Disease

In our previous reports (Kurtzberg et al. 1996; Rubinstein et al. 1998) the severity of acute GvHD was lower than is usually reported following unrelated bone marrow transplants (Sullivan et al. 1989; Szidlo et al. 1997). It was also associated with HLA-matching. Fully matched PCB transplants (6/6 matches, including high-resolution DRB1) had significantly lower rates of severe acute GvHD than did mismatched transplants (8.3% vs 24%, $P=0.031$) (Rubinstein et al. 1998). Our data as of December, 1999 include 550 engrafted patients from whom we have reports for at least 6 months post-transplant with data on acute GvHD, and 37 additional patients who reported GvHD but who did not achieve an ANC of 500. The overall correlation between the number of HLA antigens mismatched (0, 1, 2, or more) and the severity of acute GvHD (grades 0+1, 2, or 3+4) is now stronger: χ^2 (4df)=14.3, $P=0.006$,and the linear-by-linear χ^2 with 1df=12.3, $P<0.001$ (Table 2). Both the Eurocord and the Duke-Minnesota analyses (Wagner and Kurtzberg 1988; Gluckman et al. 1997; Locatelli et al. 1999) however, found no significant associations between acute GvHD and HLA.

4.2.2 Transplant-Related Events

TRE include autologous reconstitution, receiving an additional stem cell transplant, and death. The cell dose, being so important for the timing of engraftment, is necessarily among the variables related to TRE (Table 3). For this reason, we repeated the analysis of TRE using *only* patients who achieved myeloid engraftment, and there was a residual, significant association in the univariate but not in the multivariate analyses (Table 4). The age variable was, however, significant in the multivariate analysis of engrafted patients. Thus, it is possible that cell dose may only affect the probability of TRE secondarily, through its influence on the speed of myeloid reconstitution, while age may have an unrelated, independent effect.

The relevance of HLA matching to the risk of developing TRE has also been a somewhat controversial issue in PCB transplantation. These data show conclusively the association between the degree of matching (number of mismatched antigens) and the incidence of TRE. The association holds even if the fully-matched patients and those with three or more HLA antigen mismatches are excluded: those with one had significantly lower incidence of TRE than those mismatched for two ($P<0.001$). Multivariate tests using Cox logistic regression that include cell-dose, age, diagnosis, and disease stage showed HLA to be an independently significant variable, whether one considers all patients or only those who engrafted. Thus, the data indicate that HLA is certainly associated with the risk of TRE (Fig. 3).

A less studied variable that has remarkable importance in the incidence of TRE in the first 2 years post-PCB transplant is the experience of the Transplant Center with PCB transplantation (up to ten grafts vs ten or more) (Fig. 4). It should be noted that all transplant centers included had prior experience with allogeneic bone marrow transplants, and most satisfy the experience requirements for performing unrelated bone marrow transplants facilitated by the National Marrow Donor Program (NMDP). The majority of the transplants performed in non-

Table 3. Relative risk for transplant-related events (TRE)[a] during the 1st year post-transplant in all patients given unrelated placental/cord blood grafts[b]

Variable	Number of patients	Relative risk for TXP-related events (95% CI) Univariate	P	Multivariate	P
CML, FA, or SAA					
No	670	1.0		1.0	
Yes	142	1.8 (1.4–2.2)	0.001	1.5 (1.1–2.0)	0.007
Age group					
<2 years	166	1.0		1.0	
2–11 years	371	1.6 (1.2–2.1)	0.003	1.3 (0.9–1.9)	0.15
≥12 years	275	2.6 (1.9–3.7)	0.001	1.5 (0.96–2.5)	0.07
WBCs per kilogram (from initial count per $\times 10^6$)					
≥100	118	1.0		1.0	
50–99	179	1.4 (.96–2.0)	0.10	1.2 (.83–1.8)	0.29
25–49	282	1.8 (1.3–2.6)	0.001	1.4 (0.97–2.1)	0.07
	233	2.7 (2.0–3.8)	0.001	2.0 (1.3–3.0)	0.002
Number of HLA-A, -B, -DR mismatches (assigned as stated in Table 2)					
None	45	1.0		1.0	
1	314	1.6 (.95–2.7)	0.08	1.8 (1.05–3.0)	0.03
≥2	451	2.1 (1.3–3.5)	0.003	2.1 (1.3–3.5)	0.004
US and non-US transplant centers					
US	664	1.0		1.0	
Non-US	148	1.6 (1.3–2.0)	0.001	1.5 (1.1–1.8)	0.003

[a] TRE, autologous reconstitution, receipt of another graft, or death. Relapse was censored.

[b] Patients with missing data on outcome (n=50) and 2 patients who died on the day of transplant are automatically excluded from Cox regression analysis.

Table 4. Relative risk for transplant-related events (TRE)[a] during the 1st year post-transplant in patients who reached ANC ≥500

Variable	Patients (n)	Relative risk for TXP-related events (95% CI)			
		Univariate	P	Multivariate	P
CML, FA, or SAA					
No	497	1.0		1.0	
Yes	76	1.6 (1.1–2.2)	0.008	1.2 (0.8–1.7)	0.3
Age group					
<2 years	136	1.0		1.0	
2–11 years	274	1.6 (1.1–2.4)	0.017	1.6 (1.03–2.5)	0.035
≥12 years	163	2.4 (1.6–3.6)	0.001	1.9 (1.09–3.3)	0.024
WBCs per kilogram (from initial count per $\times 10^6$)					
≥ 100	107	1.0		1.0	
50–99	141	1.2 (0.7–1.8)	0.5	0.9 (0.6–1.5)	0.8
25–49	188	1.2 (0.8–1.8)	0.4	0.8 (0.5–1.3)	0.4
	137	2.0 (1.3–3.0)	0.002	1.2 (0.7–2.1)	0.5
Number of HLA-A, -B, -DR mismatches (assigned as in Table 2)					
None	38	1.0		1.0	
1	217	1.1 (0.6–2.1)	0.7	1.3 (0.7–2.4)	0.5
≥2	317	1.5 (1.1–3.6)	0.03	2.0 (1.07–3.7)	0.03
US or non-US transplant center					
US	491	1.0		1.0	
Non-US	82	1.5 (1.05–2.1)	0.026	1.4 (0.98–1.9)	0.07

[a] The transplant-related event in this analysis was death after the patient reached an ANC ≥500. Patients who had autologous reconstitution, received a backup graft, or died before reaching an ANC ≥500 were excluded from the analysis, and patients who relapsed were censored at the time of relapse.

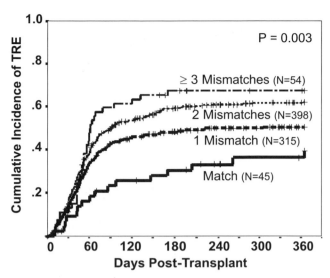

Fig. 3. HLA match level and incidence of transplant-related events

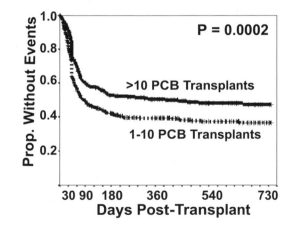

Fig. 4. Two-year incidence of TRE by transplant center experience

United States Centers were done by Centers having performed less than ten grafts, which may explain the difference in the incidence of TRE in the United States and abroad. As in 1998, there is a significant association with diagnosis: patients with aplastic anemia, Fanconi anemia or CML have higher incidence of TRE. This association in both uni- and multivariate analyses of the entire patient set could not be confirmed in the multivariate analysis of the subset of engrafted patients, suggesting that the TRE association may depend, at least in part, on the slower engraftment in patients with these diseases.

4.2.3 Leukemic Relapse

To-date, relapse has been reported in 30% (by Kaplan-Meyer analysis) of the patients who had ≥1 year of follow-up. The incidence of relapse is most strongly associated with the stage of disease at transplantation (Fig. 5), as also disclosed by earlier data in both PCB and bone marrow transplantation (Sullivan et al. 1989; Horowitz et al. 1990; Gluckman et al. 1997; Szidlo et al. 1997; Rubinstein et al. 1998; Locatelli et al. 1999). Relapse also correlates negatively with the severity of GvHD, as expected. Severe acute GvHD (grade ≥3) is associated with a probability of relapse of 9%, compared with 29% in patients without GvHD or with

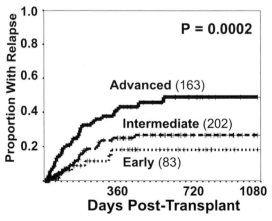

Fig. 5. Three-year incidence of relapse by stage at transplant (IBMTR criteria)

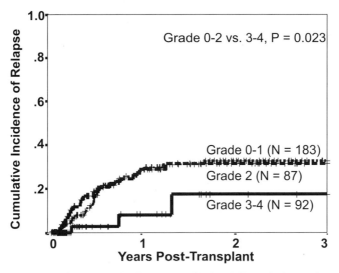

Fig. 6. Severity of GvHD and relapse rate (CML, AML, and ALL patients)

GvHD of grade ≤2 (*P*=0.023) (Fig. 6). Relapse is also less likely in the presence of chronic GvHD, whether limited or extensive, for those who survive at least 100 days (*P*=0.049). In the multivariate analysis, however, the only significant independent predictors of relapse were the stage of disease at transplantation and acute GvHD (data not shown). These data are relevant to the question of the ability of PCB transplants to provide effective graft-vs-leukemia (GvL) reactivity (Sullivan et al. 1989; Horowitz et al. 1990). It has been suggested, hypothetically, that GvL might be reduced given the lower tendency to generate severe GvHD (Linch and Brent 1989). Since the incidence of relapse is not higher than is seen after bone marrow transplant, and it is lower in association with anti-host immune reactivity of the graft, we may suggest that GvL develops well after PCB transplantation.

4.2.3.1 Event-Free Survival
The event-free survival (EFS) of patients with genetic disease is higher than in those with acquired disease (severe aplastic anemia and myelodysplasia) or leukemia. The Kaplan-Meier survival rates and 95% confidence intervals were 48% (40–55%), 29% (23–35%), and 27%

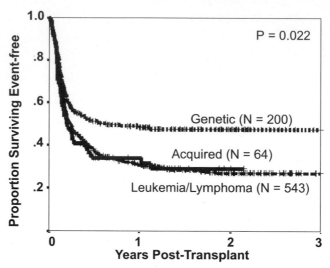

Fig. 7. Event-free survival and disease category

(23–31%), respectively, 3 years post-transplant (Fig. 7). Multivariate analysis indicates that clinical stage, cell dose and HLA are independently significant predictors of EFS in the subset of patients with ALL, AML, or CML (not shown). In the small subset of patients with genetic disease, only transplant center experience remained significant in the multivariate test. In this subset, the influence of cell dose could not be distinguished from that of patient's age, but either one in the absence of the other was a significant contributor in the multivariate test.

4.3 Conclusions

The clinical usefulness of unrelated PCB transplantation is now generally felt to be established. An increasing number of organizations collect PCB grafts for release to unrelated recipients, and an increasing number of transplant centers use PCB as one of the available sources in practical transplantation. Netcord, an international organization of cord blood banks, was established with the principal objective of improving the availability of better-matched PCB grafts (Wernet et al. 1999) to help improve the prognosis of PCB transplants and, generally, to help the development of this field. Many transplant physicians no longer consider PCB transplantation as an experimental procedure, and regulatory authorities in the United States and elsewhere feel the need to enact legal framework for the correct development of these and the speculative storage programs. A problem for the adoption of common guidelines, standards, and criteria is the relatively sparse body of results reported thus far from the recipients of related-donor grafts on which to base them. This deficit is especially serious from the viewpoint of assuring that the sets of patients analyzed are complete and that all transplants performed are reported. The absence of reporting is particularly severe in the area of speculative storage, despite the increasing commercial development of this kind of storage. For example, absolutely nothing has been reported thus far on the utilization of grafts collected on speculative grounds or on the outcomes of transplants of those units, whether autologous or to relatives whose diagnosis was made *after* the collection.

In contrast, the results of more than 1400 unrelated PCB transplants have been reported, formally or informally by the different programs. Because the numbers from individual banks are generally small and the diversity in patient- and diagnosis-dependent variables is considerable, conclusions from even this aspect of clinical PCB utilization need to be viewed as somewhat preliminary.

Overall, however, our own observations on over 900 consecutive transplants of units from the New York program indicate that the main endpoints can be related to predominantly relevant variables. Thus:

- Faster engraftment is strikingly associated with higher cell doses and, less dramatically, with HLA-A,B,DRB1-identity.
- The incidence of severe acute GvHD is associated with HLA incompatibility and patient's age and/or cell dose.
- The incidence of TRE is associated with HLA mismatches, more advanced disease and lower cell doses and/or older patient's age.
- The incidence of relapse is associated with more advanced disease, lower grades of acute GvHD and absence of chronic GvHD.

These factors and the somewhat less appreciated effects of disease diagnosis and transplant center's experience with PCB transplants are all reflected in the EFS endpoint. It is especially interesting to confirm the prognostic importance of HLA incompatibility on engraftment and acute GvHD with respect to the fully matched grafts. These deleterious effects may become more severe with increasing incompatibility, but the evidence for this is, thus far, less persuasive. On the other hand, increasing HLA incompatibility in PCB transplants brings about a parallel, progressive increase of the probability of TRE, as discussed above.

While it is now clear that PCB grafts can provide useful restoration of bone marrow function, even in the presence of HLA mismatches, the increasing evidence of the independent influence of HLA on transplant endpoints stresses the need to improve the chances of finding better HLA matches, at higher resolution. Better matches, however defined (Van Rood and Oudshhorn 1998), are particularly important for patients with genetic diseases, who do not benefit from the GvH/GvL effect of mismatched grafts in lowering relapse rates for leukemic patients. Improvement of the chances for a better match depends exclusively on increasing the overall number of fully studied PCB units available for immediate, unrestricted use. Increasing the number of collections, adopting common standards, and implementing practical coordination among academic and public PCB Banks therefore constitute paramount goals towards improving the prognosis of patients needing these transplants.

Acknowledgements. The authors are grateful to the patients worldwide who have participated in this experimental trial of the efficacy of PCB transplantation, to the transplant physicians and their centers for their support and willingness to share clinical data and to the teams of professionals responsible for the work of the program. We also recognize the financial support of the National Heart, Lung and Blood Institute, National Institutes of Health, for the period 1992–1995, and for program grants from Baxter Healthcare and CITIBANK.

References

American Academy of Pediatrics Work Group on Cord Blood Banking (1999) Cord blood banking for potential future transplantation: subject review. Pediatrics 104:116–118

Armitage S, Fehily D, Dickinson A et al (1999) Cord blood banking: volume reduction of cord blood units using a semi-automated closed system. Bone Marrow Transplant 23:505–509

Bertolini F, Lazzari L, Lauri E et al (1995) Comparative study of different procedures for the collection and banking of umbilical cord blood. J Hematother 4:29–36

Besalduch J (1985) Naturaleza y características de los precursores Granulocitico-macrofágicos en Sangre de Cordón, doctoral thesis, Universitat de València, Valencia, Spain

Broxmeyer HE, Douglas GW, Hangoc G, Cooper S, Bard J, English D, Arny M, Thomas L, Boyse EA (1989) Human umbilical cord blood as a potential source of transplantable hematopoietic stem/progenitor cells. Proc Natl Acad Sci USA 86:3828–3832

Broxmeyer HE, Hangoc G, Cooper S, Ribeiro RC, Graves V, Yoder M, Wagner J, Vadhan-Raj S, Benninger L, Rubinstein P, Braun ER (1992) Growth characteristics and expansion of human umbilical cord blood and estimation of its potential for transplantation in adults. Proc Natl Acad Sci USA 89:4109, 4114

Broxmeyer HE, Kurtzberg J, Gluckman E, Auerbach AD, Douglas G, Cooper S, Falkenburg JHF, Bard J, Boyse EA (1991) Umbilical cord blood hematopoietic stem and repopulating cells in human clinical transplantation. Blood Cells 17:313, 318

Cox DR (1972) Regression models and life tables (with discussion). J R Stat Soc (B) 34:187–200

Denning-Kendall PA, Donaldson C, Nicol AJ et al (1996) Optimal processing of human umbilical cord blood for clinical banking. Exp Hematol 24:870–874

Fauser AA, Messner HA (1978) Granuloerythropoietic colonies in human bone marrow, peripheral blood and cord blood. Blood 52:1243–1248

Ford MA, Pombo-de-Oliveira MS, McCarthy KP, Carrico KC, Vincent RF, Greaves M (1997) Monoclonal origin of T-cell malignancy in identical twins. Blood 89:281–85

Gale KB, Ford AM, Repp R, Borkhardt A, Keller C, Eden OB, Greaves MF (1997) Backtracking leukemia to birth: identification of clonotypic gene fusion sequences in neonatal blood spots. Proc Natl Acad Sci USA 94:13950–13954

Gluckman E, Broxmeyer HE, Auerbach AD, Friedman H, Douglas GW, DeVergie A, Esperou H, Thierry D, Socie G, Lehn P, Cooper S, English D, Kurtzberg J, Bard J, Boyse EA (1989) Hematopoietic reconstitution in a patient with Fanconi anemia by means of umbilical-cord blood from an HLA-identical sibling. N Engl J Med 321:1174–1178

Gluckman E, Rocha V, Boyer-Chammard A et al (1997) Outcome of cord blood transplantation from related and unrelated donors. N Engl J Med 337:373–381

Horowitz MM, Gale RP, Sondel PM, Goldman JM et al (1990) Graft-versus-leukemia reactions after bone marrow transplantation. Blood 75:555–562

Issaragrisil S (1994) Cord blood transplantation in thalassemia. Blood Cells 20:259–262

Kaplan EL, Meier P (1958) Nonparametric estimation from incomplete observations. J Am Stat Assoc 53:457–481

Kernan NA, Schroeder ML, Ciavarella D et al (1994) Umbilical cord blood infusion in a patient for correction of Wiskott-Aldrich syndrome. Blood Cells 20:245–249

Knudtzon S (1974) In vitro growth of granulocytic colonies from circulating cells in human cord blood. Blood 43:357–361

Koike K (1983) Cryopreservation of pluripotent and committed hemopoietic progenitor cells from human bone marrow and cord blood. Acta Paediatr Jpn 25:275–283

Kurtzberg J, Graham M, Casey J, Olson J, Stevens CE, Rubinstein P (1994) The use of umbilical cord blood in mismatched related and unrelated hemopoietic stem cell transplantation. Blood Cells 20:275–283

Kurtzberg J, Laughlin M, Graham ML et al (1996) Placental blood as a source of hematopoietic stem cells for transplantation into unrelated recipients. N Engl J Med 335:157–166

Linch DC, Brent L (1989) Can cord blood be used? Nature 340:676

Locatelli F, Rocha V, Chastang C et al (1999) Factors associated with outcome after cord blood transplantation in children with acute leukemia. Blood 53:3662–3671

Mahmoud HH, Ridge SA, Behm FG, Pui C-H, Fard AM, Raimondi SC, Greaves M (1995) Intrauterine monoclonal origin of neonatal concordant acute lymphoblastic leukemia in monozygotic twins. Med Pediatr Oncol 24:77–81

Nakahata T, Ogawa M (1982) Hemopoietic colony-forming cells in umbilical cord blood with extensive capability to generate mono- and multipotential hemopoietic progenitors. J Clin Invest 80:1324–1328

Prindull G, Prindull B, Meulen N (1978) Haematopoietic stem cells (CFUc) in human cord blood. Acta Paediatr Scand 67:413–416

Rebulla P., De Bernardi N, Villa A et al (1998) Evaluation of a new device for volume reduction of placental blood units by filtration through polyurethane. Vox Sang (Abstracts of the 25th Congress of ISBT) 36 (abstract)

Rowley JD (1998) Backtracking leukemia to birth. Nature Med 4:150–151

Rubinstein P (1993) Placental blood-derived hematopoietic stem cells for unrelated bone marrow reconstitution. J Hematother 2:207–210

Rubinstein P, Rosenfield RE, Adamson JW, Stevens CE (1993) Stored placental blood for unrelated bone marrow reconstitution. Blood 81:1679–1690

Rubinstein P, Taylor PE, Scaradavou A et al (1994) Unrelated placental blood for bone marrow reconstitution: organization of the placental blood program. Blood Cells 20:587–600

Rubinstein P, Dobrila L, Rosenfield RE et al (1995) Processing and cryopreservation of placental/umbilical cord blood for unrelated bone marrow reconstitution. Proc Natl Acad Sci USA 92:10119–10122

Rubinstein P, Carrier C, Scaradavou A et al (1998) Outcomes among 562 recipients of placental blood transplants from unrelated donors. N Engl J Med 339:1565–1577

SPSS Advanced Statistical Software 1999. SPSS Inc, Chicago, IL

Sullivan KM, Weiden Pl, Storb R et al (1989) Influence of acute and chronic graft-versus-host disease on relapse and survival after bone marrow transplantation from HLA-identical siblings as treatment for acute and chronic leukemia. Blood 73:1720–28

Szidlo R, Goldman JM, Klein JP et al (1997) Results of allogeneic bone marrow transplants for leukemia using donors other than HLA-identical siblings. J Clin Oncol 15:1767–1777

Tokushima Y, Sasayama N, Takahashi TA (1999) Engraftment of human cord blood stem cells separated by stem cell collection filter (SCF) in NOD/SCID mice: evidence for maintenance of SCID-repopulating activity. Blood 1 [Suppl 1]:574A

Van Rood JJ, Oudshhorn (1998) An HLA matched donor? An HLA matched donor? What do you mean by an HLA matched donor? Bone Marrow Transpl 22 [Suppl 1]:583

Vilmer E, Sterkers G, Rahimy C, Denamur E, Elion J, Broyart A, Lescoeur B, Tiercy JM, Gerota J, Blot P (1992) HLA-mismatched cord blood transplantation in a patient with advanced leukemia. Transplantation 53:1155–1159

Visser J, Alespeiti G, Stevens C, Rubinstein P (1999) Relation between numbers of CD34$^+$ cells and CFU-C in placental cord blood units. Blood 94 [Suppl 1]:4792 A

Wagner J, Kurtzberg J (1988) Unrelated donor umbilical cord blood transplantation at Duke University and at the University of Minnesota: results in 143 patients. Abstract from the 3rd Eurocord Transplant Concerted Action Workshop

Wagner JE, Broxmeyer HE, Byrd RL, Zehnbauer B, Schmeckpeper B, Shah N, Griffin C, Emanuel PD, Zuckerman KS, Cooper S, Carow C, Bias W, Santos GW (1992a) Transplantation of umbilical cord blood after myeloablative therapy: analysis of engraftment. Blood 79:1874

Wagner JE, Broxmeyer HE, Cooper S (1992b) Umbilical cord and placental blood hematopoietic stem cells: collection, cryopreservation and storage. J Hematother 1:167–173

Wagner JE, Kernan NA, Steinbuch M, Broxmeyer HE, Gluckman E (1995) Allogeneic sibling umbilical-cord blood transplantation in children with malignant and non-malignant disease. Lancet 346:214–219

Wernet P, Kogler G, Hakenberg P et al (1999) Standards and efficiency of cord blood banking by the international Netcord organisation. Blood 94 [Suppl 1]:4763A

5 Chances and Limits of Cord Blood Transplantation

E. Wunder

5.1 Introduction

Cord blood (CB) as a new source of grafting material has evoked great expectations. Because it is derived from the fetus, it contains immature cells with unusual properties: they possess higher proliferation capacity for regenerating the hematopoietic system after ablative treatment than does grafting material gained from adults, and they retain differentiation capacities for some non-hematopoietic tissues as well. The main application today is straightforward use as sibling or unrelated allograft during myoablative treatment of malignant disease. Important indications are also given in children with inherited blood cell diseases, hemoglobinopathies, immune deficiencies and other genetic diseases such as osteopetrosis and storage diseases; beside this, new perspectives for tissue replacement in the case of organ damage in adults open up, and both applications foster a family-centered strategy for CB storage.

Related and unrelated grafts are distinct immunological situations, and in the latter category CB is expected to fill gaps in conventional marrow donor banks due to its universal availability and ease of collection; here the supposed immunological immaturity is viewed as an additional advantage for graft compatibility. CB can be collected rapidly, with relatively little effort, and kept on stock in banks. Since the first demonstration of a successful CB graft in a boy with severe Fanconi's Anemia from an unaffected sibling by Gluckman and Broxmeier (1989), further detailed studies of the biological features of CB cells have been done. Considerable clinical experience has been collected meanwhile, which on the one side allows comparison with conventional grafts using bone marrow (BM) cells, as usually applied in allografts, or mobilized blood cells, the latter being the most frequently used material today, mainly for autografts; on the other side, results of related and unrelated CB grafts can be compared, and in this context, implications of the different strategies of public and family-based CB banks become apparent.

5.2 Origin and Frequency of Stem Cells in Cord Blood

The essential element of all grafting materials are immature hematopoietic cells (characterized by the surface marker CD34); after destruction of the marrow by an ablative treatment, the so-called conditioning, they regenerate hematopoiesis upon intravenous injection. While these cells occur very rarely in peripheral blood (PB) of children or adults (about ten times less frequently than in BM, unless mobilized by high-dose cytostatics and/or hematopoietic growth factors), they are found in CB at similar frequencies as in normal BM cells [1–3% of mononuclear cells (MNC)]. The site of hematopoiesis changes repeatedly in embryonal and fetal life, and towards the end of pregnancy reaches its definitive location in BM; the elevated number of stem cells in CB at birth reflects this last move from liver and spleen to the marrow space. This explains why the content at delivery shows some variation, but rarely descends to adult blood values (less than 0.1% of MNC).

5.3 Biological Properties of Cord Blood Stem Cells

Numerous studies have focussed on evaluating the quality of the hematopoietic cells in CB and compared it with those in BM cells (Mayani and Landsdorp 1998). The progenitors found in CB belong to more immature stages: the fraction of primitive burst forming units (BFU-E) prevails with 94% among all red progenitors (vs 62% in BM); the bipotent granulo-monocytic (CFU-GM) progenitor fraction comprises 29% (vs 9%) of total myeloid progenitors, and among the megakaryocytic progenitors the larger CFU-MK are more abundant in CB; finally, the less mature mixed colony progenitors (CFU-GEMM) are found up to 4 times more frequently in CB. Among the primitive progenitors, cells with high proliferative potential (HPP-CFC) have been found eight times more frequently in CB, whereas the long-term culture-initiating cells (LTC-IC) show similar frequency in both materials. After adding cytokines, they leave the dormant stage G_0 faster; it was also noted, that some colonies could form without adding cytokines in CB CD34$^+$ cells, supposedly due to internal stimulation, and both properties should give them an advantage in proliferation after grafting. Also, the phenotypically characterized CD34$^+$CD38$^-$ cells, which in mPB have been found to correlate best to graft recovery (Hénon et al. 1998), are found four times more frequently in CB than in PB. In these cells, another interesting difference has been found on the molecular level: the ends of chromosomes are composed of repetitive telomeric DNA, which limits the total proliferative potential, since cells lose 50–100 bp during each division and stop dividing once the whole telomers are used up. A telomer length of 12 kb was found in CB CD34$^+$38$^-$ cells, whereas in corresponding BM cells the telomer length was 8 kb (Vaziri et al. 1994). This once more underlines the more primitive character and higher developmental potential of CB hematopoietic cells. The "earlier" these cells are, the less they are committed; CB CD34$^+$ cells in prolonged long-term culture on feeder cells also developed early B cell progenitors (Rawlings et al. 1995), and additional experiments confirmed that bipotent, lymphomyeloid stem cells are present in CB. Finally, growth of epithelial cells in culture showed that endothelial precursors are also present (Nieda et al. 1997).

Apart from in vitro cultures, stem cells can also be assessed by injecting them into immune-deficient severe combined immunodeficient/non-obese diabetic (SCID/NOD) mice and evaluating the growth of human mature cells in the resulting chimeras. Besides erythroid, myeloid, and megakaryoid cells, the human CB stem cells also produced lymphoid cells, and most recently Civin's group demonstrated that CB contains eightfold more human SCID long-term engrafting cells than mobilized adult blood CD34+ cells (Leung et al. 1999; see also Wang et al. 1997).

5.4 Clinical Aspects of Cord Blood Grafting

While these experimental results indicate the very high biological quality of CB cells as grafting material, it is also clear that the amount of cells that can be obtained during delivery is limited. The complete CB is therefore frozen after collection during delivery, adding the protecting agent dimethyl-sulfoxide (DMSO), usually without further manipulation that would imply loss of hematopoietic cells. Experiences from BM and mPB indicate that the number of hematopoietic cells needed for rapid recovery of neutrophils in blood after grafting is related to the body weight of the patient and has a threshold value (To 1994). Therefore, in the first attempts at CB grafting, only smaller children were treated; with growing confidence in CB grafts, adults are now also receiving these grafts more frequently. However, today no firm definition of minimal requirements or optimal range of total cell number in CB grafts exists, and recommendations of different grafting centers vary markedly. We will come back to this point later.

Grafting with conventional materials has been developed in separate ways, depending on the origin of the graft. In the earliest approach, which continues to be practiced, BM is taken from an optimally human leukocyte antigen (HLA)-matched donor, who is either related or unrelated. The graft not only transfers regenerating hematopoietic stem cells from the donor, but also its immune system; the therapeutic intention here is immunological eradication of residual malignant cells in the patient that resisted prior cytostatic treatment.

The later and alternative approach is to employ the patient's own BM cells, freeze them, treat the patient with high-dose chemotherapy and/or radiation, and inject the defrosted cells. Here the main intention of the graft is to apply the highest doses of chemotherapy despite their myelotoxicity, and to restore hematopoiesis afterwards. This method avoids acute and chronic graft-versus-host disease (GvHD),which is an often severe and not rarely lethal complication of allografting. On the other hand, the recurrence risk of the malignancy is higher in most diseases due to the lack of the "graft versus malignancy" effect or just to a compromised immune system and possible regrowth of residual malignant cells. As for their biological properties, CB can be applied in the same settings as BM or mPB: as related or unrelated allograft, and also possibly as autograft, if the child is genetically healthy.

Use of CB grafts of a healthy newborn for grafting a sibling who carries a severe Mendelian disease concerns few families, but is usually their best chance for making normal life of such an individual possible. Besides defects in blood and immune cell production, diseases such as osteopetrosis can be treated, where dysfunctional osteoclasts (derived from monoblasts) are replaced. Also, metabolic diseases such as thesaurismoses (M. Gaucher, Pfaundler-Hurler's disease, etc.) are successfully treated by a sibling graft, and the usefulness of storing healthy CB in such targeted families is undebated. The possibility of prenatal diagnosis in a growing number of these diseases gives further support to CB transplantation.

Another group of individuals with a genetic load may benefit from CB stored for future use within the family: those who carry an increased risk of developing malignant tumors. In future, special care for these families may include banking of CB of all newborn children; if a malignancy occurs, CB from an unaffected and completely or partly matched sibling could help. If in future gene replacement therapy becomes practicable, an optimal choice will be to correct the defect in own stem cells and make an autograft.

The main group of patients who profit today from CB banks are those with manifest malignant diseases. If such an individual needs a graft, a decision has to be made concerning which kind of graft would be best, and if it should be an allograft, the question arises as to whether a suitable donor is available.

In the conventional allograft setting, the risk of severe and chronic GvHD is lowest if the donor of BM or mPB is related and perfectly HLA-matched. However, often – especially in older patients – there is no matched sibling available, and then an unrelated donor must be found. Because of the extraordinary variety of the HLA types, the chance of finding a matching unrelated donor is only given if very numerous voluntary donors have been typed; in spite of international networking between respective data banks, it is often difficult to find a donor, especially if the patient belongs to an underrepresented ethnic minority. If a possible donor is identified, he often cannot be localized due to address changes, etc., and even if the search for a donor is successful, usually 4–6 months elapse until the graft is available. There- fore, CB banks which can relatively rapidly collect the necessary high numbers of grafts appear as an attractive alternative: CB is collected at deliveries, instead of being thrown away, and adjacently it is typed and frozen. In case of need in the future, it is immediately available. In addition, with the notion of less mature immune cells in CB, less complications with GvHD are to be expected, and even HLA mis- matches are supposed to be tolerable with less complications.

5.5 Results of Systematic Cord Blood Collection

In Europe and the United States, several CB banks have been founded. They are organized as either public or private banks. Public banks collect CB samples for use on demand as unrelated allogeneic grafts. Part of the running costs are covered by payments at handing out of a suitable CB batch. The biggest is New York Cord, and the total number of such grafts collected by April 2000 in the United States and Europe plus Israel was 442,263 (International BM Donor Registry, Leiden). The alternative are private banks, where CB grafts are stored on demand and at the expense of the client families; in this case the grafts remain the property of the donor and are reserved for use within the family as related allograft or even autologous graft in case of need. The total number of batches collected so far is estimated to be 70,000 (Lampeter and Egger 2000); their minimum time of conservation is usually guaran- teed for 10 years and can be extended further on demand.

While CB for public banks is collected routinely in delivery clinics with the simple consent of the mothers, collection for private banks needs more personal effort. It is worth mentioning that initially, sponta- neous active interest was noted in pregnant women who had experi- enced previous transplantation in a family member or friend, or who had health-related professions and thus realized the value of a matching graft

Table 1. Results of nationwide cord blood (*CB*) collection for central banking

	Mean	Maximum	Minimum
Processing time (*n*=475) (delivery–storage)	19 h (median)	31.5 h	4 h
CB volume (*n*=370)	79±27 ml	210 ml	30 ml
NC count (*n*=300)	$6.6\pm4.6\times10^8$	31.0×10^8	0.5×10^8

Table 2. Technical and administrative problems encountered in cord blood (*CB*) collection

Small CB volume (<50 ml)	14%
Microbial contamination	3,3%
Delay in transportation (>24 h)	<4%
Incomplete accompanying forms	6%
Incorrect packaging	1%

being immediately available. More recently, the majority of interested mothers is motivated by the media and by obstetricians. If CB is collected in clinics that are not located in the vicinity of CB banks, as is the case in family-based CB banks, training of the midwives has to be extended accordingly. In a convenient collection system, which is provided by the CB bank, the clamped umbilical cord is punctured sterilely, and the blood is collected in a closed bag containing anticoagulants; rapid and reliable transportation of the bag must be guaranteed. After arrival in the CB bank, volume, cell number, and sterility are tested, and after mixing the blood with DMSO in a storage bag, conservation is started in a controlled freezing apparatus, which compensates for temperature jumps at phase transitions.

As an example, average yields of such collection in a large series of consecutive CB collections in a private European bank are given in Table 1. Besides occasional collection of small CB volumes, serious complications such as microbial contamination, where the sample has to be discarded, were rare (Table 2). The amount of collected volumes and cells is important with regard to applicability in adults, and therefore the frequency distribution in the series is displayed in Figs. 1 and 2. While the total number of nucleated cells varies between 2×10^8 and 2×10^9, it is most frequently between 5 and 10×10^8.

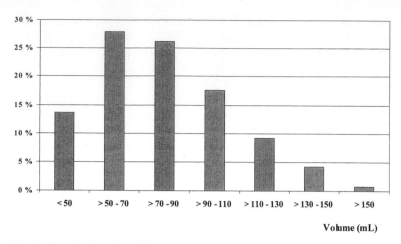

Fig. 1. Distribution of cord blood volume (*n*=400)

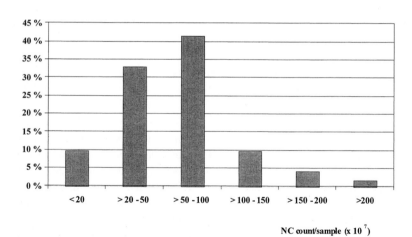

Fig. 2. Distribution of nucleated cell (*NC*) count (*n*=300)

Recommendations concerning the minimal number of cells needed per kilogram of body weight given by different authors and applied in the respective centers vary largely. The lowest rate considered admissible is 0.1×10^7 nucleated cells/kg in Minnesota (Wagner et al. 1996), whereas it is 0.7×10^7 nucleated cells/kg in New York (Rubinstein et al. 1998); in Düsseldorf, 1×10^7 nucleated cells/kg are indicated (Lampeter and Egger 2000). This implies that a priori no shortage of cells should be expected for younger children as receivers; at older age and higher weight, suboptimal or insufficient grafting may result. In the CB collection shown here as an example, a maximum weight in adults is calculated as 66 kg for the average collected cell number if the most conservative of the above-mentioned recommendations is applied. It has to be considered, however, that with higher cell numbers per kilogram, better grafts result, as will be detailed below. There is a great interest in collecting as many cells as possible, and since it is the case that the obtained volume of CB increases with more practice of the midwives, efforts for sufficient training are important in order to obtain the best yields possible.

5.6 Clinical Outcome in Related and Unrelated Cord Blood Grafts

Besides the American series (see Chap. 4, this volume), the results of 143 related and unrelated CB grafts in Europe, generated by 45 centers until early 1997 and assembled as the Eurocord series, were published initially (Gluckman et al. 1997) and 543 have now been evaluated (Gluckman et al. 1999). The duration of aplasia was identical in both related and unrelated grafts, and on average lasted around 30 days; this is markedly longer than in autologous grafts of BM (18.5 days) and mAB (12 days), provided that optimal quality grafts are given with regard to the content of GM-progenitors or CD34$^+$ cells. Fourteen percent of all patients had not engrafted by day 60. Overall platelet recovery at 56 days appears to be about double as long as in conventional autografts, while in allografts recoveries are also slower. In 38% of related CB grafts and 61% of unrelated CB grafts, platelets had not recovered after day 60. In both situations, higher cell numbers per kilogram body weight and HLA identity were clearly favorable.

With regard to overall survival at 1 year, recipients of CB from related donors did much better: 63% of them were alive, whereas only 29% of unrelated CB recipients were still alive (both groups were composed of similar diseases and patients).HLA identity played a crucial role in related CB grafts as well for 1-year survival (73%, vs 33% in mismatches) and for acute GvHD (9% vs 50%, $P<0.001$), whereas there was no significant difference in unrelated grafts.

Relapses encountered in leukemias after grafting were somewhat more frequent in related (10/38) than in unrelated grafts (7/41),whereas total lethal events, including other complications such as interstitial pneumonitis, veno-occlusive disease, GvHD, etc., were lower in related (30 of 78) than in unrelated grafts (38 of 65).

An interesting trend of grafted cells emerged from these data when results were grouped into patients who had received more than 3.7×10^6 cells/kg, and those who had received less: in unrelated CB-grafts 1-year survival was 41% vs 22% ($P=0.07$), whereas it was higher and the difference was less pronounced in related grafts with 57% vs 68%. Also, duration of aplasia and platelet recovery was reduced to 35 days with higher cell dose, while it was 53 days with less in related, and 25 and 47 days in unrelated transplants. In another report, after grafting a median of 3×10^7 cells/kg of unrelated CB and application of G-CSF, a median aplasia time of 22 days was achieved (Kurtzberg et al. 1996).

5.7 Conclusion

Taking today's experiences together, CB as grafting material has definite advantages concerning the superior capacities of the hematopoietic cells being contained. Also, in the allograft setting, immaturity of the immune cells leads to less frequent and less severe stages of GvHD than comparable allografts of adult BM or mobilized blood, but this complication remains a serious risk. Overall, results such as for 1-year survival are far better after related than after unrelated CB grafts, and perfect HLA matching is highly preferable, at least in related CB grafts; this has been confirmed in the most recent update of a threefold larger series of European related and unrelated grafts than in the first Eurocord evaluation. With regard to all recovery kinetics and engraftment data, the results are best if high numbers of cells per kilogram are applied. Since the number of CB cells usually obtained is appropriate for smaller children, they currently profit best from CB grafts, whereas with increasing body weight recovery gets slower for neutrophils and especially platelets, and the risk of infectious complications mounts.

This limitation could be overcome by stem cell expansion. Recent highly interesting data have been related, which suggests that this may be possible soon (Douay and McNiece 2000): Douay achieved 20-fold amplification of LTC-IC after in vitro culture of CB cells in a gas-permeable bag for 14 days, using the cytokine cocktail indicated by Piacibello et al. (1997). McNiece divided the CB and expanded one part using different growth factors during 10-day incubation; adults were then grafted with those cells combined with the untreated cells. Fast recoveries of both platelets and neutrophils resulted, virtually without limit by body weight. It thus can be expected that in the near future CB grafts for adults will be more frequently applied than today, where their main role is to complement conventional allografting in case no matching adult donor is at hand. In malignancies occurring later in life, autologous CB stem cells will offer a choice with less risks, since it avoids immunological complications and usually contains no tumor cells. For any use, best efforts are necessary to enhance the yield of CB harvests at deliveries.

In view of developing new biotechnologies, it is of great interest to look forward to additional forthcoming applications of CB beyond today's therapies in tissue engineering: already, endothelial stem cells have been found, which may be useful for the repair of damaged blood vessels, for example, in coronary or other arterial disease. Also, other stem cells may be detected in CB and be used for restoring other organs such as bone or cartilage. Since autologous stem cells are ideal for tissue replacement, this option is an additional reason to store CB individually in family-based banks.

Acknowledgements. I would like to thank Dr. Lampeter and Dr. Egger (Leipzig, Germany) for letting me use data from their cord blood collection.

References

Douay, McNiece (2000) Contributions at the international conference on haematopoietic stem cell biology and transplantation, Paris, 25–29 March 2000

Gluckman E, Broxmeyer HE, Auerbach AD et al (1989) Hematopoietic reconstitution in a patient with Fanconi's anemia by means of umbilical-cord blood from an HLA-identical sibling. N Engl J Med 321:1174–1178

Gluckman E, Rocha V, Boyer-Chammard A et al (1997) Outcome of cordblood transplantation from related and unrelated donors. N Engl J Med 337 (6):373–381

Gluckman E, Rocha V, Chastang CL (1999) Umbilical cord blood hematopoietic stem cell transplantation. Eurocord-Cord Blood Transplant Group. Cancer Treat Res 101:79–96

Hénon Ph, Sovalat H, Becker M, Arkam Y, Ojeda-Oribe M, Raidot JP, Husseini F, Wunder E, Bouderont D, Audhuy B (1998) Primordial role of CD34+38-cells in early and late trilineage haemopoietic engraftment after autologous blood cell transplantation. Br J Haematol 103:568–581

Kögler G, Somville T, Gobel U, Hackenberg P, Knipper A, Fischer J, Adams O, Krempe C, McKenzi C, Ruttgers H, Meier W, Bellmann O, Streng H, Ring A, Rosseck U, Rocha V, Wernet P (1999) Haematopoietic transplant potential of unrelated and related cord blood: the first six years of the Eurocord/Netcord Bank Germany. Klin Padiatr 211(4):224–232

Kurtzberg J, Laughlin M, Graham ML et al (1996) Placental blood as a source of hematopoietic stem cells for transplantation into unrelated recipients. N Engl J Med 335:157–166

Lampeter EF, Egger D (2000) Storage of cord blood for related or unrelated use. In: Schultze W (ed) High-dose therapy and transplantation of hematopoietic cells. Blackwell Wissenschafts-Verlag, Berlin

Leung W, Ramirez M, Civin CI (1999) Quantity and quality of engrafting cells in cord blood and autologous mobilized peripheral blood. Biol Blood Marrow Transplant 5(2):69–76

Mayani H, Landsdorp PM (1998) Biology of human umbilical cord blood-derived hematopoietic stem/progenitor cells. Stem Cells 16:153–165

Nieda M, Nicol A, Denning-Kendall P et al (1997) Endothelial cell precursors are normal components of human umbilical cord blood. Br J Haematol 98:775–777

Piacibello W, Sanavio F, Garetto L et al (1997) Extensive amplification and self renewal of human primitive hematopoietic stem cells from cord blood. Blood 89(8):2644–2653

Rawlings DJ, Quan SG, Kato RM et al (1995) Long-term culture system for selective growth of human B-cell progenitors. Proc Natl Acad Sci USA 92:1570–1574

Rubinstein P,Carrier C, Scaradavou A et al (1998) Outcomes among 562 recipients of placental-blood transplants from unrelated donors. N Engl J Med
339(22):1565–1567
To LB (1994) Establishment of a clinical threshold cell dose: correlation between CFU-GM and duration of aplasia. In: Wunder E, Sovalat H, Henon
PR, Serke S (eds) Hematopoietic stem cells. Mulhouse Manual Alpha Med
Press, Dayton OH
Vaziri H, Dragowska W, Allsopp RC, Thomas TE, Harley CB, Lansdorp PM
(1994) Evidence for a mitotic clock in human hematopoietic stem cells: loss
of telomeric DNA with age. Proc Natl Acad Sci USA 91:9857–9860
Wagner JE, Rosenthal J, Sweetman R et al (1996) Successful transplantation
of HLA-matched and HLA-mismatched umbilical cord blood from unrelated donors: analysis of engraftment and acute graft-versus-host disease.
Blood 795–802
Wang JCY, Doedens M, Dick JE (1997) Primitive human hematopoietic cells
are enriched in cord blood compared with adult bone marrow or mobilized
peripheral blood as measured by the quantitative in vivo SCID-repopulating
cell assay. Blood 89:3919–3924

6 Legal and Ethical Issues Involved in Cord Blood Transplantation and Banking

K. Seelmann

Quite frequently, medical progress leads to new questions and solutions in the legal field as well. Even when at first glance it is merely a question of applying tried and agreed legal principles to new facts, the principles themselves do not always remain quite untouched. The debate on the regulation of reproductive medicine, for example, had an influence on the definition of a person in law, and also on the constitutional principle of human dignity. The debate on euthanasia has redefined the position of human life in the hierarchy of values. Towards the provisional end of the debate in Germany, questions of legal autonomy and legally protected reverence appear in a new light.

With this experience in mind, one is not surprised to note that it soon becomes apparent that the question of legal issues involved in cord blood transplantation and banking does not simply imply the application of non-controversial principles to new cases. This becomes evident in the similarities of the medical novelty of collection and banking of cord blood with the well-known practice of blood donation on the one hand, since blood is involved, but also with spinal bone marrow donation on the other hand, since stem cells are involved, and here trouble begins, because different rules apply. The proposal to apply the rules of blood donation as the more closely related parallel (Annas 1999, p. 1521) leaves aside certain specific aspects. Newborn children are involved – this also is a new situation. Furthermore, something is collected that is not part of a human being, that would in any case be disposed of, and that until now, in principle, was of no interest to anybody. The latter aspect distinguishes cord blood collection from blood donation as well as from spinal bone marrow transplantation. Another difference between blood donation and spinal bone marrow donation is the fact that cord blood cannot be regenerated. In view of the absolutely fundamental legal difficulties resulting from these specificities, we shall concentrate on central legal issues. Thus, we shall first deal with rights in cord blood (6.1), and we shall then look at issues raised by consent to collection, banking, and transplantation of cord blood (6.2). Obviously, the second issue – consent – depends on the answer to the first, i.e., who has what rights in cord blood? And finally, we have to look at the legal possibilities and limitations of banking (6.3), themselves dependent on the rights in blood and the requirements for collection, banking, and transplantation. The third set of issues concerning banking may influence answers to the first two issues, in as much as the objective in respect to the ways of banking may have an influence on the right in the blood and on the requirements of consent.

6.1 Rights in Cord Blood

When we ask the question of the rights in cord blood, the most comprehensive right, the right of ownership, first comes to mind. It is widely accepted that any part of the human body not separated from the body cannot reasonably constitute an object of ownership rights (Taupitz 1992, p. 1091). To be considered, albeit in part, as an object, contradicts the rights of a person to a subjectivity in the emphatic sense. In this sense, a human being cannot be an object to himself/herself either, that is why in the legal sense he/she has no right of ownership of himself/herself.

As long as cord blood circulates between placenta and fetus or neonate, it is owned by nobody, not by the pregnant woman either, of course. The inability to be owned ends when a part of the body becomes separated from it (Brohm 1998, p. 199; Taupitz 1992, p. 1092). This point is reached at the latest the moment cord blood is collected. The legal fate of this part of the body can be determined by different theoretical approaches, which in practice usually all lead to the same result.

It can be argued that in the instant of separation from the body, the right of ownership of the person is constituted of which the part was part of, consequently, the right to transfer ownership to others devolves upon this person (Heinrichs 1999, § 90 N 3; Dilcher 1995, § 90 N 16). It goes without saying that this person can simply relinquish his/her rights without contractual transfer; as a consequence, somebody else can appropriate the ownerless object.

It was formerly argued that in the instant of separation, a part of the body becomes ownerless, in other words continues to be ownerless (Coing 1957, § 90 N 4). The person to whom the part was connected has the first right of acquisition, he/she can decide whether to become owner of the severed part of the body. If he/she decides in favor of acquisition, this part is at his/her disposal, e.g., to be sold in a sales contract – he/she can also simply give up ownership. The latter case leads to the situation that would arise without his/her acquisition of the part, i.e., a third party is allowed to appropriate this part of the body.

The traditional disposal of the placenta and the umbilical cord including cord blood (Annas 1999, p. 1521) has to be interpreted in law as the waiving of ownership rights in the cord blood on behalf of the neonate by its guardian (Burgio and Locatelli 1997, p. 1164; Annas 1999, p. 1522) (view 1), or as the waiving of the right of acquisition of the ownerless blood (view 2), respectively. In both cases, the consequences are the same: the hospital can make free use of the cord blood. If it decides to dispose of it, no further legal issues arise. The situation seems slightly more complicated when the placenta is used for cosmetic purposes. As a person's rights have precedence over ownership rights – as we shall see in detail – the sale of the placenta, at least with this objective and combined with the aim to make a profit, by the hospital or any person would not have been allowed without consent (Schröder and Taupitz 1991, p. 71–83).

What changes does the new situation bring about with respect to the former transfer of ownership in cord blood to the hospital? From the point of view of legal ownership, nothing changes. In the case of storage in favor of the donor, however, the ownership rights are not waived or the first right of acquisition is exercised, in either case the consequences are the same: ownership does not change, if no action is taken (view 1), or if ownership is established by the guardians (view 2). Co-ownership by the hospital or the blood bank based on the processing of the blood (§ 950 BGB) may be negated, both in view of the paramount value of cord blood compared with the outlay for collection and processing, and in view of the processing in favor of the owner. Potential co-ownership of the mother to the extent of the part of the mother's cells in the cord blood can be assumed. This is, however, of no consequence, since it has to be surmised that the mother in her declaration of intent in her capacity of one of the guardians of the neonate – if she should decide to act in addition in her own name – decides on both counts in parallel.

The issue of ownership becomes more complex when the guardians want to transfer ownership to a public blood bank in favor of third persons. Theoretically, it is possible to assume that this is a case of transfer of ownership in favor of this institution in the name of the neonate (view 1) or a case of renouncing the first right of acquisition in the name of the newborn (view 2). It is debatable whether this intended new ownership can be constituted, since the guardians are expected to keep their mind on "the welfare of the child" (§ 1627 BGB), according to family law. In the case of normal blood donation by the child to third persons, it is generally assumed that this can hardly ever contribute to "the welfare of the child" (Kern 1981, p. 739; Schmidt-Elsaesser 1987, p. 167). A similar assumption applies to the donation of spinal bone marrow. Why has action in these two cases to be regarded fundamentally opposed to the welfare of the child? A closer look soon reveals that in either case not only a transfer of ownership or waiving of the right of ownership is involved, which in other cases, especially in the case of service in return, is not necessarily in conflict with the welfare of the child: both parallel cases, the donation of blood and of spinal bone marrow, involve an operation which is not in the interest of the child itself. Such an operation requires consent, and in those instances where it is given by a guardian, its scope is limited to decisions for the welfare of the child. We will, therefore, look at cord blood donation from this angle and determine whether in this case consent is necessary and what conditions, if any at all, are attached to consent in the case of cord blood donation.

6.2 Consent

The first question is: does cord blood donation constitute an infringement of bodily integrity which would require consent? Based on the definition of blood as "discarded tissues", this is sometimes considered doubtful (Lind 1994, p. 829; Gluckman et al. 1998, p. S69).

If cord blood is collected before the cord is cut, the operation is performed on an integral part of the body. There is, however, a specific aspect to it: this part is shortly to be severed from the body in any case, and this operation does not have any negative consequences on the physical well-being. If cord blood is collected after the cord has been cut, this is not in relation with a living body or with a functioning part of the body. This differentiation of collection before or after cutting the umbilical cord is, however, of no consequence. In both cases, the question of infringement of physical integrity has to be answered in the affirmative, albeit for another reason.

At any rate, in German administration of justice a very extensive definition of the body applies to the case of legal protection of physical integrity. Based on certain ethical concepts, the human body is considered "a sphere of existence and of determination", materialized only in corporeal existence (BGHZ 124, 52, 54). In consequence of this stance, the principles of protection of the individual apply, which entail, above all, the definition of the scope of physical integrity by self-determination. Self-determination encompasses not only the body itself, but also parts and substances severed from it.

German administration of justice draws the conclusion that severed parts intended to be reunited with the body – such as in the case of autologous blood donation – still form a functional unit with the body from the legal point of view, and, consequently, that their destruction or damage to them is considered as bodily harm (BGHZ 124, 52, 55). In German administration of justice, the rights of the individual precede rights of ownership also in severed parts intended for a third party (BGHZ 124, 52, 55). In consequence of this, the right of disposal of the donor remains intact even after transfer of ownership, comparable to the right of disposal in the case of letters sent and transferred, the publication of which, though they are owned by the recipient, can be prevented, based on the rights of the individual (Taupitz 1992, p. 1093). In the case of allogous blood donation, the use of this blood against the explicitly or tacitly expressed will of the donor can be grounds for claims of compensation. This may also mean that the permission to use blood from allogous donations may be withdrawn at any time. Expenditures arising for a third party, resulting from this change of mind in the donor, will have to be met by the donor who changed his/her mind and thus caused the costs to a third party.

What are the consequences of all this for the need of consent to the collection of cord blood? There is no question that the collection of cord blood, whether before or after the cutting of the umbilical cord or even independent of this, whether it is a case of autologous or allogous donation, is an infringement of physical integrity as defined by the rights of the individual. Effective consent is a prerequisite.

What legal requirements does this consent have to meet, who can give it and what limitations are there to it?

Persons under age are legally represented by their parents, that is to say by their father *and* mother (§ 1626 I BGB). If the ability to understand the issue is lacking, the legal representatives can give consent (Jescheck and Weigend 1996, p. 382; Köhler 1997, p. 254). Principally, this is considered cumulative not alternative (Burgio and Locatelli 1997, p. 1163: "informed consent of the parents"). Today's position of Eurocord is based on the concept of informed consent of the newborn parents or at least of the mother (Fernandez 1998, p. 585). The guidelines of the Federal Association of Physicians (Bundesärztekammer) define the consent of the father as "desirable" (Bundesärztekammer 1999 A-1301). In American publications, the exclusive right to consent of the pregnant woman or mother is assumed (Annas 1999, p. 1522; Sugarman et al. 1997, p. 942). In ordinary daily matters, German law assumes that the parent who decides has been given power of attorney for this decision by the other parent (Diederichsen 1999, § 1627 N 1, § 1356). In important matters comprising big elective medical operations, however, the consent of both legal representatives is indispensable. In this respect, the collection of cord blood comes to lie in a gray area. As a rule, fathers tend to endorse the pertaining decisions by the pregnant mothers. In view of the possibly considerable consequences for the child, the decision in the case of autologous donation may duly be considered important and comparable to that with respect to an important therapeutic intervention. The collection of cord blood would in this case require the consent of both parents as guardians. This, of course, only holds good in as far and as long as autologous cord blood donation is considered to be of potential therapeutic use. It does not hold good when public banking is the only medically sensible option.

It has further to be decided whether this consent of the guardians has objective limitations. In the discussion of ownership issues, we have already mentioned that legal representation of minors by their guardians has to be guided by the welfare of the child (§ 1627 BGB). The welfare of the child clearly is a limiting factor in the exercise of representation. Bearing the therapeutic potential in mind, autologous donation cannot possibly be considered as not in the interest of the child's welfare. Allogous donation intended for public banking, on the other hand, raises legal issues of some importance.

In the case of normal blood or bone marrow donation, there is general agreement in pertaining publications: whether the guardians' consent to collection in order to save the life of a person the child closely relates to, especially in the case of a near relative, affords cover for this action is, however, a controversial issue. It is sometimes (Hirsch 1994, pp. 959, 960; Dumoulin 1998, p. 86; Schoeller 1994, p. 93) argued that it may indirectly be in the interest of the child to save these persons, also to refute possible later reproaches addressed to the guardians (Schöning 1996, pp. 218–220). *E contrario*, the operation itself and its direct consequences for the minor can be given more weight and the admission of consent be denied despite such special circumstances (Kern 1981, p. 740; Schmidt-Elsaesser 1987, pp. 170–173). Though it was different before (Lenckner 1960 p. 460), today it is commonly acknowledged in publications on the subject that allogous donation of blood or bone marrow in favor of persons outside the inner social circle of the donor is not in the interest of the child's welfare, and the guardian's consent is, therefore, inadmissible (Schoeller 1994, p. 94; Kern 1981, p. 740; Schöning 1996, p. 218).

Is this also valid for allogous cord blood donation, i.e., donation in favor of a public bank? This does not per force follow. Considering the ratio legis., the differences between normal blood or bone marrow donation on the one hand and cord blood donation on the other hand are considerable. Cord blood donation does not involve the child's organism (Burgio and Locatelli 1997, p. 1165). If any relevance with respect to physical integrity can be seen, then it can only relate to self-determination. To deprive the newborn of this right – contrary to physical integrity – cannot infringe the welfare of the child. Physical integrity in the narrow sense might be involved – under certain conditions and in the future – by refusing autologous donation. If this holds good, refusal of any cord blood collection would constitute an injury by omission to the child, for which consent is inadmissible. Among other things, this would imply that autologous cord blood donation were part of lex artis in medicine – which, at least until now, is not the case – and, furthermore, that there are no reasonable medical arguments against it – which is not the case either. If, nowadays, physical integrity in the narrow sense is not involved by any decision with respect to cord blood collection, consent to collection for storage in a public bank cannot be detrimental to the welfare of the child, since it does not involve the child to a greater extent than the decision for disposal of the cord blood. If the guardians consent to it, there is no infringement of physical integrity in the wider sense, as in the case of collection without their consent. There is no infringement of physical integrity in the narrow sense at all, even if the guardians do not consent. Parents have to weigh decisions and measures that will, with certainty or at least high probability be harmful to the child, against other less harmful measures (Peschel-Gutzeit 1997, § 1627 N 19).

6.3 Banking

Finally, we come to legal issues with respect to banking. Private banking, i.e., autologous cord blood donation and storage for use by the donor or close relatives, raises a number of legal and ethical aspects: human dignity as protected by the constitution, social equity, and the constitutional principle of the welfare state, and finally, issues of commercial law (competition and truth in advertising).

Human dignity would be in jeopardy, if in the context of cord blood collection the newborn were disparaged as a means for other purposes. This, however, does not hold good in every case of allogous donation in favor of another member of the family, since this does not simply make use of the neonate. Things would look different if a child were begot for the sole purpose of saving another living child. Such motivation for begetting, however, could in any case not be legally examined (Burgio and Locatelli 1997, p. 1164). That this motivation is a general possibility cannot make the individual case of cord blood collection appear in an unfavorable light from the constitutional point of view. It is altogether a different question, whether the system of private banking is liable to be the cause of such motivation and action of this kind to any relevant extent or encourage such action. Even if this is assumed, the danger of misuse is in all probability no greater than in the case of bone marrow donation, which, from the constitutional point of view, is nonetheless considered unproblematic.

Private banking, however, raises the serious issues of justice, equity, and the welfare character of the state. The question of social justice and equity exists as long as autologous blood donation on the one hand makes sense medically, on the other hand, it is out of reach for financial reasons for many patients because of the high cost of collection and storage (Sugarman et al. 1995 p. 1785, 1997 p. 941).

Nowadays, most states indirectly or directly acknowledge the right to health care, which offers minimal care for everybody and can be compared with the level of subsistence guaranteed by general consent. States where this demand is met by a health care insurance system have to guarantee sufficient insurance coverage for adequate medical treatment. If and when private banking is part of the state of the art in medicine, steps in this direction have to be taken. Even under these circumstances, private banking could not be forbidden, but it would have to be made financially accessible to everybody. While banking is not medical standard (yet?), private banking cannot be forbidden by law.

Competition law may limit private banking (Ratzel 1998, pp. 514–515). Under certain circumstances these limits may jeopardize the concept of private banking. Advertising pharmaceuticals, if not directed at health care personnel, is restricted by German competition law in order to protect the consumers. Above all, making use of fear and advertising with respect to tumors is not allowed (§§ 11, lit 7, 12 lit 1 HWG – Gesetz über die Werbung auf dem Gebiet des Heilwesens). This might also severely limit advertising private cord blood banking to the general public. Individual information to pregnant women mentioning the danger of leukemia for the yet unborn child and the positive therapeutic effects of stem cells taken from the child's cord blood is not allowed. It would not be allowed, even if the positive effect of stem cells from one's own cord blood should be unequivocally proved. As long as this remains an assumption or in the face of serious doubt, advertising the positive effects would not be allowed, because the claim would not be true. According to the opinion of the Federal Association of Physicians (Bundesärztekammer), there is no medical indication known as yet (Bundesärztekammer 1999 A-1300). The simple application of the clauses limiting competition might already lead to a complete ban on advertising private banking to the general public, and this would, of course, include direct mailings to pregnant women.

Public banking, the last aspect we will turn to, also raises complex legal issues. They include the protection of potential recipients from life-threatening illnesses, which means there has to be a link between the stem cells and the data of the donor (Sugarman et al. 1997, pp. 939–940). If an illness of the donor is diagnosed after the collection only, or if the need for examination of the donor for illnesses or infectious diseases that were unknown at the time of collection or for which diagnosis has become possible after collection arises, it must be possible to identify the donor, at least further data of the donor have to be made available. Ideally, the donor would regularly provide an update. The link between the stem cells and the personal data has an impact on the donor's self-determination in respect of disclosing information. There is in this case a conflict between the right of the potential recipient to a safe product and the right of the donor to autonomy in disclosing information.

Explicit consent to this link between the data and the stem cells, let alone a regular update, does not solve the problem, since consent has again to be given by the legal representatives who decide at the expense of the child and therefore against its welfare. The possibility to withdraw consent at any time, which would have to be stipulated until the stem cells are used to treat a third party, does not in any way defuse the situation. Until the child reaches maturity, the guardians only could withdraw consent.

A potential infringement of the negative freedom of information poses a further difficulty (Hellermann 1993, pp. 117–129; 147–148, 163–164, 184ff; Schulze-Fielitz 1996 p. 371 N 63), because the link between data and stem cells makes it possible to disclose information about previously unknown illnesses or infectious diseases of the donors or other information detrimental to their interest. This might be in their interest, but if they do not want to disclose this information, this can also infringe their negative right of freedom. Considering the ambiguity of this assessment, the guardians will have to be conceded the authority to define the welfare of the child to the best of their knowledge in this situation.

There remains the diminution of the right of self-determination in respect to information by the link between personal data of the donor and the stem cells. A radical solution would be not to collect personal data at all, but to put the processed stem cells under quarantine for a reasonably long time instead. There is a medical disadvantage to this solution: in the mean time, the stem cells cannot be used in therapy at a time when they seem particularly effective. To avoid this, one would have to forgo quarantine, even without data collection, and leave the risk to be decided upon by the recipient. This solution, however, also has its disadvantages, as third parties might be at the risk of infection. We have, therefore, to consider a solution in which the indispensable link between data and stem cells is created in such a way as to stand up to legal examination even in cases where this link does not work in the interest of the donor as well (if the legal representatives have made their interest in important information reaching the donor clear). Data could be anonymized (Burgio and Locatelli 1997, p. 1165). This would have the advantage, that certain tests, e.g., on a blood sample, could still be carried out, but also the disadvantage that no further data would be available from the donor. Whether separate storage of data and (processed) blood, strengthened by appropriate firewalls (Sugarman et al. 1997, p. 940), could allay the legal reservations, is not easily decided. If it is assured that on the basis of the stored data, no further individual enquiries, e.g., after bone marrow donation, are made to the donor or his/her relatives, and the data are stored in case of dire need, e.g., to protect the recipient from life-threatening dangers, consent by the guardians to the storage of the donor's data may be considered as just within the limits of what can be asked of a person in the way of a citizen's solidarity.

To end this brief survey of some central legal and ethical issues with respect to *banking* and based on it, we will once more give our attention to the issue of consent. The question has to be asked, to what extent the validity of consent to blood collection depends on the information about the various types of banking and their advantages and disadvantages. In the case of medical treatment, a physician must point out the different existing therapies to render a patient's consent effective, and generally speaking, the requirements this information has to meet are higher the more easily treatment can be postponed or can be done without (Tröndle and Fischer 1999, pp. 1236–1238). For purely prophylactic measures, such as cord blood collection, this is particularly pertinent. Both varieties of banking and, of course, the decisive difference – that in the case of private banking the cord blood is later available to the donor – have to be pointed out. The controversial points in the therapeutic benefit of private banking have to be explained as well.

Here, we come to a close to briefly recall the central issues, which to me seem to be in particular need of discussion:

Whereas constructional differences in the issue of ownership are of little consequence, acceptance of physical integrity, interpreted in the light of individual rights, however, raises considerable problems. The question of the position of the child's father in the consent, and above all, the question of the welfare of the child as criteria for collection for public banking have yet to be answered. I, personally, consider consent by the father necessary; consent does, however, not infringe the welfare of the child. The difficulties of private banking are connected with the concept of "truth in advertising", public banking is in difficulties with respect to the protection of data. The latter, to my mind, can be overcome. Information to the public about the different ways of banking is indispensable, if banking is offered at all.

References

Annas GJ (1999) Waste and longing – the legal status of placental-blood banking. N Engl J Med 340:1521–1524

Brohm W (1998) Forum: Humanbiotechnik, Eigentum und Menschenwürde. JuS 38: 197–205

Bundesärztekammer (1999), Richtlinien zur Transplantation von Stammzellen aus Nabelschnurblut (CB = Cord Blood). Dtsch Ärztebl 96:A-1297–A-1304

Burgio GR, Locatelli F (1997) Transplant of bone marrow and cord blood hematopoietic stem cells in pediatric practice, revisited according to the fundamental principles of bioethics. Bone Marrow Transplant 19:1163–1168

Coing H (1957) Comment on § 90 BGB. In: von Staudinger J (ed) Kommentar zum Bürgerlichen Gesetzbuch, 11th edn. De Gruyter, Berlin

Diederichsen U (1999) Comment on §§ 1297ss BGB. In: Palandt H (ed) Bürgerliches Gesetzbuch. Beck, Munich

Dilcher H (1995) Comment on § 90 BGB. In: von Staudinger J (ed) Kommentar zum Bürgerlichen Gesetzbuch, 13th edn. Sellier-de Gruyter, Berlin

Dumoulin JF (1998) Organtransplantation in der Schweiz – das Recht am Scheideweg zwischen Leben und Tod. Stämpfli, Bern

Fernandez MN (1998) Eurocord position on ethical and legal issues involved in cord blood transplantation. Bone Marrow Transplant 22 [Suppl 1]:584–585

Gluckman E, Rocha V, Chastang C (1998) Cord blood banking and transplant in Europe. Bone Marrow Transplant 22 [Suppl 1]:S68–S74

Hellermann J (1993) Die sogenannte negative Seite der Freiheitsrechte. Duncker und Humblot, Berlin

Heinrichs H (1999) Comment on §§ 1ss BGB. In: Palandt H (ed) Bürgerliches Gesetzbuch. Beck, Munich

Hirsch G (1994) Fortschritte der Medizin – Herausforderung an das Recht. In: Letzgus K (ed) Für Recht und Staat – Festschrift Herbert Helmrich. Beck, Munich, pp 953–965

Jescheck HH, Weigend T (1996) Lehrbuch des Strafrechts Allgemeiner Teil, 5th edn. Duncker und Humblot, Berlin

Kern BR (1981) Der Minderjährige als Blutspender. FamRZ 28:738–740

Köhler M (1997) Strafrecht Allgemeiner Teil. Springer, Berlin Heidelberg New York

Lenckner T (1960) Die Einwilligung Minderjähriger und deren gesetzlicher Vertreter. ZStW 72:446–463

Lind SE (1994) Ethical considerations related to the collection and distribution of cord blood stem cells for transplantation to reconstitute hematopoietic function. Transfusion 34:828–833

Peschel-Gutzeit LM (1997) Comment on § 1627 BGB. In: von Staudinger Kommentar zum Bürgerlichen Gesetzbuch, 12th edn. Sellier-de Gruyter, Berlin

Ratzel R (1998) Juristische Stellungnahme zum Angebot der Firma VITA 34 zur Konservierung von Nabelschnurblut. Frauenarzt 39:514–515

Schmidt-Elsaesser E (1987) Medizinische Forschung an Kindern und Geisteskranken – zur Strafbarkeit von Forschungseingriffen an Einwilligungsunfähigen. Metzner, Frankfurt a.M.

Schoeller B (1994) Vorschlag für eine gesetzliche Regelung der Organspende vom lebenden Spender. Lang, Frankfurt/Main

Schöning R (1996) Rechtliche Aspekte der Organtransplantation unter besonderer Berücksichtigung des Strafrechts. Schulthess, Zürich

Schröder M. Taupitz J (1991) Menschliches Blut verwendbar nach Belieben des Arztes? Enke, Stuttgart

Schulze-Fielitz H (1996) Erläuterungen zu Art. 5 GG N. 63. In: Dreier H (ed) Grundgesetz – Kommentar. Mohr Siebeck, Tübingen

Sugarman J, Reisner EG, Kurtzberg J (1995) Ethical aspects of banking placental blood for transplantation. JAMA 274:1783–1785

Sugarman J, Kaalund V, Kodish E, Marshall MF, Reisner EG, Wilfond BS, Wolpe PR (1997) Ethical issues in umbilical cord blood banking. JAMA 278:938–943

Taupitz J (1992) Privatrechtliche Rechtspositionen um die Genomanalyse: Eigentum, Persönlichkeit, Leistung. JZ 47: 1089–1099

Tröndle H, Fischer F (1999) Strafgesetzbuch und Nebengesetze – Kommentar. Beck, Munich

7 Fetal Somatic Gene Therapy – A Preventive Approach to the Treatment of Genetic Disease: The Case For

C. Coutelle, M. Themis, H. Schneider, T. Kiserud, T. Cook,
A.-M. Douar, M. Hanson, A. Pavirani, C. Rodeck

7.1 Why Fetal Somatic Gene Therapy?

With the Human Genome Project reaching its goal to determine the complete sequence of the human genome, questions concerning the use of this information for the improvement of health and in particular for the prevention of human disease are shifting more and more to the centre of attention. Obviously, an increasing number of DNA sequences with so far unknown function will be linked to disease phenotypes and will give rise to novel diagnostic and therapeutic strategies. The genetic background of many monogenetic diseases will finally be discovered, and for many other conditions, genetic factors involved in particular diseases, often manifesting only later in life, will be revealed. Therapeutic approaches will in many cases be based on the development of novel drugs, but progress made in the field of gene delivery and expression will also lead to effective gene therapy strategies using the DNA itself as the therapeutic agent. It can be predicted that the continuously broadening array of genome-based diagnostic and predictive tests will lead to increasing demands for preventive measures to avoid diseases rather than having to treat such conditions.

Since gene therapy offers in many instances, and particularly for inherited genetic diseases, a causative approach to treatment, it could also become the ideal means for prevention of such diseases. However, although many of these conditions show disease manifestation already at or shortly after birth, most present efforts towards their treatment by somatic gene therapy have concentrated on adult patients. Several years ago, we suggested that it may be possible to avoid the development of severe manifestations of early-onset genetic diseases by applying gene therapy vectors in utero (Coutelle et al. 1995; Douar et al. 1996). We also hypothesised that this may allow the targeting of otherwise inaccessible organs or tissues and still expanding stem cell populations and thus provide a basis for permanent somatic genetic correction. Gene application to the fetus may also avoid immune-sensitisation and, thereby, facilitate repeated treatment after birth. Furthermore, we suggested that, depending on issues such as reliability, safety and technical logistics, gene therapy in utero could also provide an alternative therapeutic option to decision-making following prenatal diagnosis and could, therefore, significantly alter attitudes to antenatal screening programmes for genetic diseases.

In parallel with work carried out by several laboratories in the United States, our team is investigating the feasibility of in utero gene delivery on different animal models. The aims of these investigations are to define diseases for which fetal gene therapy may be beneficial; to find the best suitable animal models, routes of vector application and optimal times during gestation for gene delivery to certain organ systems; to investigate different vector systems for their efficiency in fetal gene transfer and expression; and to identify the risks and ethical problems which may arise from this novel approach to somatic gene therapy.

7.2 Which Diseases?

Initially, fetal gene delivery will be restricted to well-defined monogenic disorders which are caused by the absence or inactivation of an essential gene product that does not require fine gene regulation for its expression. In the first instance, diseases presenting early in life would have priority. However, if somatic gene therapy in utero turns out to be efficient and safe, it may become a preventive approach to the treatment of less severe conditions which still have a serious impact on life expectancy such as the heterozygosity for LDL-receptor mutations.

Most groups that have attempted fetal gene transfer in animals so far have related these studies to cystic fibrosis (CF). This is not astonishing, since CF is the most common severe autosomal recessive disease with an incidence of 1 in 2000 newborns in our population. It affects about 5000 individuals in Britain (Jackson 1989). CF is caused by mutations in the gene coding for the cystic fibrosis transmembrane regulator (CFTR) protein. It is characterised by early onset and systemic manifestation affecting most exocrine organ systems and in particular the lung, pancreas, gut and biliary tract, leading to high morbidity and shortened life expectancy. The CFTR protein is detected during normal human embryogenesis already at 7 weeks of gestation in the yolk sac, and shortly afterwards in ciliated tracheal cells. It is strongly expressed in the intestine at 12 weeks and has been detected at 24–25 weeks in the apical domain of ciliated airway epithelia and in the collecting ducts of the submucosal glands of the airways (Gaillard et al. 1994). This suggests that it may play an important physiological role in human fetal development. It is broadly acknowledged that prevention or delay of irreversible organ damage in particular of the airway epithelia may be achieved by early treatment of cystic fibrosis.

Another disease with pulmonary manifestation is deficiency in surfactant protein B, a perinatal condition affecting the expansion of the neonatal lungs after birth. This could be treated by prenatal delivery of the gene encoding this protein to the lungs but, different to cystic fibrosis, the gene vector would have to target the alveoli rather than the airway epithelia.

Several metabolic conditions with perinatal presentation such as phenylketonuria or OTC deficiency, the different storage diseases or deficiencies in vital serum proteins such as the haemophilias may also be amenable to gene therapy directed to the fetal liver. Genetic disorders of the skin, e.g. epidermolysis bullosa or ichthyosis may be particularly accessible to treatment in utero, and the skin may also serve as an organ for bioproduction of systematically required proteins such as blood-clotting factors or erythropoietin.

Application to neurological and neuro-muscular genetic diseases such as spinal muscular dystrophy and DMD, which are characterised by progressive muscle atrophy and death within the first few years of life, is more difficult to address due to anatomical inaccessibility, but the fetal approach may offer some advantages, provided the right developmental window is found. Early-onset haematological diseases such as the rare ADA deficiency, or the much more frequent haemoglobinopathies are strong candidates, but may also be approached by in utero cell transplantation, with or without genetic manipulation.

7.3 Which Vectors?

The single application of a physiologically regulated therapeutic gene introduced into all relevant organs and remaining active throughout life without potential side effects would be the ideal strategy for fetal somatic gene therapy. Although none of the present vector systems meets all of these criteria, different vector systems offer certain beneficial features applicable for this approach. Because of their efficiency or their potential for long-term gene expression, viral vectors appear presently best suited for in utero gene therapy.

7.3.1 Adenovirus Vectors

Adenoviral vectors have been applied in most investigations on fetal gene delivery in different animal systems because of their high efficiency in gene transfer and expression. This makes them particularly useful for the exploration of the different approaches to fetal gene delivery (Holzinger et al. 1995; McCray et al. 1995; Sekhon and Larson 1995; Vincent et al. 1995; Baldwin et al. 1997; Douar et al. 1997; Woo et al. 1997; Wang et al. 1998; Iwamoto et al. 1999; Lipshutz et al. 1999a; Schachtner et al. 1999; Schneider et al. 1999; Themis et al. 1999; Yang et al. 1999a). However, conventional adenoviral vectors suffer from only transient expression and show problems in repeated delivery due to immune reactions. It will, therefore, be interesting to find out if the immune reactions can be avoided in fetal life or if they can be reduced by application of the novel "gutless" adenovirus vector systems (Schiedner et al. 1998).

7.3.2 Retroviral Vectors

Retroviral vectors are able to integrate into the host genome and could, therefore, serve as vector systems for permanent gene delivery. Retroviruses require dividing cells for infection and are, therefore, not likely candidates for in vivo gene transfer into fully differentiated or slowly regenerating organs during postnatal and adult life. In the fetus, the non-synchronised cell division and the relatively short half-life of the virus appear to cause problems, as demonstrated by the low efficiency of retroviral infection of the fetus (Hatzoglou et al. 1995; Pitt et al. 1995; Douar et al. 1997; Porada et al. 1998). This may, however, be overcome with very high virus titres, as reported for infection of adult hepatocyte in vivo by D. Jolly (American Society of Gene Therapy Meeting, June 1999; Corporate Symposium), by the introduction of producer cells in vivo (Douar et al. 1997) or by the application of lentiviral vectors, a group of retroviral vectors which do not require cell division for entry into the nucleus (Miyoshi et al. 1998).

Efforts are also made to combine the high infectibility of adenovirus with the integrative capacity of retroviruses by construction of adeno-retrovirus chimera systems (Feng et al. 1997; Ramsey et al. 1998).

7.3.3 Adeno-associated Virus

Adeno-associated virus (AAV) is obviously a very interesting and promising novel vector system to achieve longer-lasting gene correction (Wang et al. 1999), in particular since no human pathology has been associated with this virus. However the mechanism for its long-term persistence and expression is not completely understood, it is only

capable of incorporation of up to 4 kb into its genome and its production
is still rather cumbersome (Linden and Woo 1999).

7.4 Which Animal Models?

Fetal gene delivery has been investigated in mice (Holzinger et al. 1995;
Baldwin et al. 1997; Douar et al. 1997; Woo 1997; Lipshutz et al. 1999a;
Schachtner et al. 1999; Schneider et al. 1999; Türkay et al. 1999), rats
(Hathoglou et al. 1995; Sekhon and Larson 1995) rabbits (Wang et al.
1998) and sheep (Holzinger et al. 1995; McCray et al. 1995; Pitt et al.
1995; Vincent et al. 1995; Porada et al. 1998; Iwamoto et al. 1999;
Themis et al. 1999; Yang et al. 1999a).

Mice offer the great advantage of ease of maintenance and breeding
and the availability of a range of mouse models for human genetic
disease. However, their small size and differences in physiology limit
their usefulness with respect to procedures applicable in the human
fetus.

Other animals may offer a certain advantage in size, besides the
availability of specific animal models such as the Watanabe rabbit
(LDLR deficiency) or haemophilia dogs.

We have found the sheep fetus to be particularly well suited for in
utero gene delivery as it is a well-established animal model relevant to
human fetal physiology. It has a consistent gestation period and provides
predominantly singleton pregnancies. It also has a good tolerance to in
utero manipulations, has anatomical features allowing interventions
which are applicable to the human fetus and shows important similarities to humans in the development of the immune system. In addition,
screening systems are underway in New Zealand to find spontaneous CF
heterozygote mutations in sheep which could be bred to produce a
CF-sheep (Tebbutt et al. 1996). There is also an interest in generating a
CF-sheep by nuclear transplantation/cloning procedures (Harris 1997).

7.5 How to Apply?

7.5.1 Intra-amniotic Application

Gene delivery by injection into the intra-amniotic cavity is possible for all the discussed animal models. In rodents, it requires laparotomy (Papaioannou 1990), but it can be performed transcutaneously with ultrasound guidance in larger animals such as the sheep. A major limitation of intra-amniotic vector delivery is its dilution by the relatively large volume of the amniotic fluid. In addition, without specific organ targeting, a large portion of the vector will primarily reach the fetal skin and the amniotic membranes. The amniotic fluid is swallowed by the fetus, reabsorbed by the gastro-intestinal tract and then excreted through the kidneys (Brace 1995). The flow of liquid is usually from the airways into the amniotic cavity. However, the fetus also displays episodes of breathing movements (Badalian et al. 1994), leading to an influx of amniotic fluid into the lungs. We and others have observed marker gene expression in the lungs, the intestinal system and the skin after intra-uterine vector administration (Holzinger et al. 1995; McCray et al. 1995; Pitt et al. 1995; Sekhon and Larson 1995; Vincent et al. 1995; Douar et al. 1997; Iwamoto et al. 1999). Recently we have also used the widespread infection of the fetal skin and the amniotic membranes for the production of human factor IX in mice, which reached therapeutic values in utero and was still detectable after birth (Schneider et al. 1999).

7.5.2 Systemic Application Through the Fetal Circulation

Access to the fetal circulation in rodents can be achieved by intraplacental injection (Woo et al. 1997), by intracardial application (Wang et al. 1998) or by injection into the yolk sac vessels (Schachtner et al. 1999). In sheep, adenoviral vector delivery into the umbilical vein has been achieved at 60 days of pregnancy (term 147) by fetoscopy after maternal laparotomy and externalisation of the umbilical vein (Yang et al. 1999a). We used an ultrasound-guided minimally invasive transcutaneous technique at and after day 102 of gestation to deliver adenoviral vectors carrying the β-galactosidase marker gene and human factor IX cDNA, respectively, to the sheep fetus (Themis et al. 1999). This has resulted in about 30% of cells staining positive for β-galactosidase in the liver. Somewhat less but significant staining was also observed in the adrenal gland cortex. This high level of hepatocyte infection can in part be attributed to the peculiarities of the fetal circulation, which directs about 50% of the placental blood flow directly into the liver and from there to the vena cava. The remaining 50% placental blood bypasses the liver through the ductus venosus and mixes in the vena cava inferior with the blood coming from the liver, before reaching the heart. As expected, very little transgene expression was observed in the lung, as this organ is, to a large extent, short-circuited by the foramen ovale and the ductus arteriosus. However, of all the tissues served by the general circulation only the adrenal gland cortex showed substantial β-galactosidase gene expression. The well-known increase in the adrenal blood flow during stress of up to 200–300% (Jensen and Berger 1991) may well have contributed to a high exposure of this organ to the vector. In addition, a so far unknown tropism of the adenovirus for the adrenal cortex seems to exist and could perhaps be exploited for gene therapy of disorders of this endocrine gland such as congenital adrenal hypoplasia.

That most other organ systems did not show significant transgene expression on the protein level compared with the liver and adrenal gland may be due to virus dilution in the circulation. However, nested polymerase chain reaction (PCR) and reverse transcription PCR (RT-PCR) analysis indicated that a substantial amount of the vector reached the general circulation and was at least transcribed in most organs including the gonads (Themis et al. 1999).

Using the factor IX adenovirus construct we achieved therapeutic factor IX plasma levels, presumably mainly from infected hepatocytes, in fetuses and newborn lambs, as determined by ELISA, indicating the potential therapeutic usefulness of this minimally invasive gene delivery technique to the liver (Themis et al. 1999).

7.5.3 Other Routes of Application

Gene delivery to the fetal airways by intra-amniotic delivery is usually fairly poor. In sheep, quite elaborate surgical techniques on the open uterus have been employed for direct instillation of retroviral (Pitt et al.

1995) and adenoviral vectors (McCray et al. 1995; Iwamoto et al. 1999) into the trachea of fetal lambs. In both cases, however, transgene expression was not very effective. More recently, fetoscopy after laparotomy has been applied for intra-tracheal injection of adenovirus, which, however, led predominantly to alveolar infection (Sylvester et al. 1997; Yang et al. 1999a). Delivery to the fetal airways is obviously a difficult task, as already experienced postnatally.

Intraperitoneal application of adenoviral vectors has been used to infect multiple tissues and in particular hepatocytes in rodents (Hatzoglou et al. 1995; Lipshutz et al. 1999b). Intraperitoneally injected retroviral vectors have led to infection of haematopoietic stem cells (Porada et al. 1998). Direct hepatic injection has also been applied for infection of hepatocytes (Lipshutz et al. 1999a).

Persistent gene expression for 16 months was observed after intramuscular application of a β-galactosidase expressing adenovirus in fetal mice at day 14–15 of gestation (Yang et al. 1999b).

7.6 When to Apply?

The preferred gestation time for gene application will depend in many cases on the normal expression pattern of the particular gene during fetal development. This may be different for different genes and will also be strongly influenced by technical considerations and the morphological and physiological development of the fetus. For delivery to the airways, for instance, one may want to make use of the breathing movements, which start in the mouse at day 15 and in humans around the 10th week of gestation. In most cases, however, two general objectives apply which may determine the choice of time in fetal development for therapeutic gene delivery: (1) to avoid germline transmission and (2) to circumvent the induction of an immune response to vector and transgene products.

7.6.1 Avoiding Germline Transmission

Fetal somatic gene therapy attempts not to modify the genetic content of the germline. After compartmentalisation of the primordial germ cells (PGCs) in the gonads, it seems unlikely that the transgene would reach these cells by other means than perhaps through the circulation. In the mouse, PGCs are first distinguishable as a distinct population in the extra-embryonic mesoderm of the posterior amniotic fold at the beginning of the 7th day of gestation. They become incorporated into the base of the allantois, where they form a cluster, and then start to migrate to the genital ridge at about the 8th day of gestation. The first PGCs reach the genital ridge by day of gestation 11.0–11.5, and the process is completed by full colonisation of the gonad primordium on day 13 (Hogan et al. 1994). This process is similar in humans, where the PGCs are fully compartmentalised by the 7th week of gestation (Gillman 1948).

After application of adenovirus vectors to the fetal circulation in sheep, we observed infection of the gonads by PCR, but could not detect any gene expression in these organs by immunohistochemistry or RT-PCR (Themis et al. 1999). Similarly, infection of the fetal gonads was also detected by PCR after intraperitoneal application of retroviral vectors to fetal sheep, but no germline transmission could be detected by PCR analysis of sperm derived from three rams after in utero treatment (Porada et al. 1998). Obviously, more analyses are required to come to final conclusions on this issue.

7.6.2 Avoiding Immune-Competence

One of the major theoretical advantages of fetal gene therapy is the possibility of avoiding immune reactions of the fetus against vector or transgene product, especially since immune reactions have turned out to be one of the major problems in adult somatic gene therapy using adenovirus vectors (Coutelle 1997).

Whether or not the human fetus will develop an immune response against a vector and/or the transgene product is most likely to be highly dependent on the stage of fetal development at which it is administered. The human immune system is acquired progressively throughout the first half of pregnancy, by maturation of the fetal immune cell repertoire in the thymus, which allows self-recognition, and by the development of a complete humoral and cellular immune response against foreign antigens. Although the immune system of the human fetus begins to develop relatively early in pregnancy, it does not completely fulfil its tasks until a year after birth, when normal adult immune-competence functions are achieved. Since the immune-response in sheep develops similarly to that of humans by mid-gestation (Morris 1986), sheep may be a suitable model to investigate the immune reactions to vector and transgene. Yang et al. could not find neutralising antibodies after applying 10^{11} adenoviral particles into the umbilical vein of sheep fetuses at 60 days of gestation. Even a repeat administration of 1×10^{12} particles did not cause any reaction at this time of gestation, while a substantial antibody titre was observed at application of 10^{11} particles on day 125 of gestation. The authors, therefore, conclude that the sheep fetus is immune-naive up to some stage between days 90 and 125 of gestation. However, adenovirus application at early gestation does not induce immune tolerance at later stages (Yang et al. 1999a). A similar conclusion was also reached by Iwamoto after application of up to 10^{11} pfu adenovirus to the airways of fetal sheep around day 60 (Iwamoto et al. 1999).

7.7 Which Ethical Issues Need to Be Considered?

Fetal somatic gene therapy is presently a hopeful but still purely experimental approach to somatic gene therapy based on animal studies. Nevertheless, since this is a novel approach to gene therapy, it is appropriate to consider the possible ethical implications that it may have in respect to future application in humans. The issue of fetal somatic gene therapy was already raised by the British Clothier Committee on the Ethics of Gene Therapy and was seen to pose no special ethical objections if used for the treatment of a serious disease in an affected individual, as long as germline transmission could be avoided (Clothier 1992). In the United States, the ethical issues involved have been outlined and discussed on balance in favour of this approach (Fletcher and Richter 1996). A recent preproposal by French Anderson to the National Institute of Health Recombinant Advisory Committee (NIH RAC) to conduct human fetal gene therapy trials within 2–3 years has met with some controversy (Couzin 1998; Billings 1999; Schneider et al. 1999), which has also led to a more cautious attitude of the British GEMAC (Gene Therapy Advisory Committee 1999). The main concern is inadvertent germline transmission, although Anderson and colleagues have shown this to be very unlikely in former experiments (Porada et al. 1998). The

RAC called for more experiments to assess both the risks of the proposed protocols and their chance of success and suggested that long-
term studies on in utero gene transfer in sheep and many generations of
mice should be conducted and that different diseases should be considered as candidates (Couzin 1998).

This is certainly the right way forward, but it should be underlined
that inadvertent germline transmission is a general point to consider in
gene therapy and is not specific to fetal gene therapy. In this context it is
important to note that the calculated frequency of naturally occurring
endogenous insertional mutations in humans of about 1 in 10 individuals is substantially higher than the suggested upper tolerable limit of 1
insertion event per 6000 sperm due to a gene delivery protocol (Kazazian 1999). Based on these calculations, the FDA has recently decided to
allow certain clinical phase 1 trials to go ahead despite incomplete
biodistribution data or detection of gonadal presence of the vector
(Epstein et al. 1999). Furthermore, gene therapy it is not the only
iatrogenic procedure which carries the risk of germline modification,
but unlike the undirected germline damage caused, for instance, by
high-dose chemotherapy, the introduction of a single normal gene sequence can be detected relatively easily over the endogenous mutated
variant. However, so far all studies on germline transmission of gene
therapy vectors administered in adults or in utero have failed to demonstrate transmission into the germline. Even if it turns out that germline
transmission cannot be completely avoided, the ethical questions arising
will be centred around a benefit/risk analysis and will have to be judged
in relation to the spontaneous occurrence of germline mutations and to
other therapeutic procedures which cause germline damage.

Much more obvious potential risks arise from the issue of safety and
reliability of the proposed procedures. Similar to most obstetric interventions, they concern the mother as well as the fetus, with a bias for life
and well-being clearly in preference of the mother. This raises the
question of reliability, which will obviously be required with more
stringency for in utero gene therapy than is acceptable for postnatal
approaches. Since termination is a reasonably safe maternal option for
dealing with an inherited genetic disease, any in utero gene therapy will
be expected to be reliable in avoiding this disease and not to cause
additional damage. During the introductory phase of transferring this
technology to humans, this danger may not be easily ascertained and

will require particular care with respect to informed maternal consent based on detailed counselling and the understanding of risks versus benefits.

Once proven to be safe and reliable, other considerations such as cost/effectiveness and equal accessibility as well as its relation to genetic screening, and in particular prenatal screening, will have to be addressed. We are fully aware that for this approach to gene therapy to become truly preventive it will require the broad support and ethical acceptance of the majority of the population. This can only be achieved on the basis of intensive experimental research and in-depth information about its benefits and risks in discussions with health professionals and the lay public.

Acknowledgements. This work has been supported by grants from the Leopold Muller Bequest/CF-Trust, the Sir Jules Thorn Charitable Trust and SPARKS. We wish to thank Dr. S.Adebakin, Ms I. Hopton-Scott, Ms. S. Jezzard, Prof. D. Noakes, Ms S. Buckley, and Dr. S. Waddington for their helpful contributions to this work.

References

Badalian SS, Fox HE, Chao CR, Timor-Tritsch IE, Stolar CJH (1994) Fetal breathing characteristics and postnatal outcome in cases of congenital diaphragmatic hernia. Am J Obstet Gynnecol 171:970–975

Baldwin H, Mickanin C, Buck C (1997) Adenovirus-mediated gene transfer during initial organogenesis in the mammalian embryo is promoter-dependent and tissue-specific. Gene Ther 4:1142–1149

Billings PR (1999) In utero gene therapy. The case against. Nat Med 5:255–256

Brace R (1995) Progress toward understanding the regulation of amniotic fluid volume: water and solute fluxes in and through the fetal membranes. Placenta 16:1–18

Clothier CM (1992) Report of the Committee on the Ethics of Gene Therapy. HMSO, London

Coutelle C (1997) Gene therapy for cystic fibrosis – strategies, problems and perspectives. In: Strauss M, Barranger JA (eds) Concepts in gene therapy. de Gruyter, Berlin, pp 313–343

Coutelle C, Douar A-M, College WH, Froster U (1995) The challenge of fetal gene therapy. Nat Med 1:864–866

Couzin J (1998) RAC confronts in utero gene therapy proposal. Science 282:27

Douar A-M, Themis M, Coutelle C (1996) Fetal somatic gene therapy. Hum Mol Reprod 2:633–641

Douar AM, Adebakin S, Themis M, Pavirani A, Cook T, Coutelle C (1997) Foetal gene delivery in mice by intra-amniotic administration of retroviral producer cells and adenovirus. Gene Ther 4:883–890

Epstein S, Bauer,S, Miller,A, Pilaro, A, Noguchi P (1999) FDA comments on phase I clinical trails without vector biodistribution data. Nat Genet 22:326

Feng M, Jackson WH, Goldman C.K, Rancourt C, Wang M, Dusing SK, Seigal G, Curiel DT (1997) Stable in vivo gene transduction via a novel adenoviral/retroviral chimeric vector. Nat Biotechnol 15:866–870

Fletcher J, Richter G (1996) Human fetal gene therapy: moral and ethical questions. Hum Gene Ther 7:1605–1614

Gaillard D, Ruocco S, Lallemand A, Dallemans W, Hinnaraski J, Puchelle E (1994) Immunochemical localization of cystic fibrosis transmembrane conductance regulator in human fetal airways and digestive mucosa. Pediatr Res 36:137–143

Gene Therapy Advisory Committee (1999) Report on the potential uses of gene therapy in utero. Health Departments of the United Kingdom, Nov 1998. Hum Gene Ther 10:689–692

Gillman J (1948) The development of the gonads in man, with a consideration of the role of fetal endocrines and histogenesis of ovarian tumours. Contrib Embryol 32:81–92

Harris A (1997) Towards an ovine model of cystic fibrosis. Hum Mol Genet 6:2191–2194

Hatzoglou M, Moorman A, Lamers W (1995) Persistent expression of genes transferred in the fetal rat liver via retroviruses. Somat Cell Mol Biol 2:265–278

Hogan B, Beddington R, Constantini F, Lacy E (1994) Manipulating the mouse embryo: a laboratory manual, 2nd edn. CSH Press, USA, pp 19–113

Holzinger A, Trapnell BC, Weaver TE, Whitsett JA, Iwamoto HS (1995) Intraamniotic administration of an adenovirus vector for gene transfer to fetal sheep and mouse tissue. Pediatr Res 38:844–850

Iwamoto H, Trapnell BC, McConnell CJ, Daugherty C, Whitsett JA (1999) Pulmonary inflammation associated with repeated prenatal exposure to an E1, E3-deleted adenoviral vector in sheep. Gene Ther 6:98–106

Jackson AD M (1989) The natural history of cystic fibrosis. In: Goodfellow P (ed) Cystic fibrosis. Oxford University Press, Oxford, pp 1–11

Jensen A, Berger R (1991) Fetal circulatory responses to oxygen lack. J Dev Physiol 16:181–207

Kazazian H (1999) An estimated frequency of endogenous insertional mutations in humans. Nat Genet 22:130

Linden R, Woo SLC (1999) AAVant-gard gene therapy. Nat Med 5:21–22

Lipshutz G, Flebbe-Rehwaldt L, Gaensler KM (1999a) Adenovirus-mediated gene transfer in the midgestation fetal mouse. J Surg Res 84:150–156

Lipshutz G, Flebbe-Rehwaldt L, Gaensler KM (1999b) Adenovirus-mediated gene transfer to the peritoneum and hepatic parenchyma of fetal mice in utero. Surgery 126:171–177

McCray PB, Armstrong K, Zabner J, Miller DW, Koretzky GA, Couture L, Robillard JE, Smith AE, Welsh MJ (1995) Adenoviral-mediated gene transfer to fetal pulmonary epithelia in vitro and in vivo. J Clin Invest 95:2620–2632

Miyoshi H, Blomer U, Takahashi M, Gage FH, Verma IM (1998) Development of a self-inactivating lentivirus vector. J Virol 72:8150–8157

Morris B (1986) The ontogeny and comportment of lymphoid cells in fetal and neonatal sheep. Immunol Rev 91:219–233

Papaioannou VE (1990) In utero manipulations. In: Copp AJ, Cockroft DL (eds) Postimplantation mammalian embryos. A practical approach. IRL Press, Oxford, Washington, pp 61–80

Pitt BR, Schwarz MA, Pilewski JM, Nakayama D, Mueller GM, Robbins PD, Watkins SA, Albertine KH, Bland RD (1995) Retrovirus-mediated gene transfer in lungs of living fetal sheep. Gene Ther 2:344–350

Porada C, Tran N, Eglitis M, Moen RC, Troutman L, Flake AW, Zhao Y, Anderson WF, Zanjani ED (1998) In utero gene therapy: transfer and long-term expression of the bacterial neo(r) gene in sheep after direct injection of retroviral vectors into preimmune fetuses. Hum Gene Ther 9:1571–1585

Ramsey W, Caplen NJ, Li Q, Higginbotham JN, Shah M, Blaese RM (1998) Adenovirus vectors as transcomplementing templates for the production of replication defective retroviral vectors. Biochem Biophys Res Commun 246:912–919

Schachtner S, Buck CA, Bergelson J, Baldwin M (1999) Temporally regulated expression patterns following in utero adenovirus-mediated gene transfer. Gene Ther 6:1249–1257

Schiedner G, Morral N, Parks RJ, Wu Y, Koopmans SC, Langston C, Graham FL, Beaudet AL, Kochanek S (1998) Genomic DNA transfer with a high-capacity adenovirus vector results in improved in vivo gene expression and decreased toxicity. Nat Genet 18:180–183

Schneider H, Coutelle C (1999) In untero gene therapy. The case for. Nat Med 5:256–257

Schneider H, Adebakin S, Themis M, Cook T, Pavirani A, Coutelle C (1999) Therapeutic concentrations of human factor IX in mice after gene delivery into the amniotic cavity: a model for the prenatal treatment of haemophilia B. J Gene Med 1:424–432

Sekhon HS, Larson JE (1995) In utero gene transfer into the pulmonary epithelium. Nat Med 1:1201–1203

Sylvester K, Yang EY, Cass DL, Crombleholme TM, Adzick NS (1997) Fetoscopic gene therapy for congenital lung disease. J Pediatr Surg 1(32):964–969

Tebbutt SJ, Harris A, Hill DΓ (1996) An ovine CFTR variant as a putative cystic fibrosis causing mutation. J Med Genet 33:623–624

Themis M, Schneider H, Kiserud T, Cook T, Adebakin S, Jezzard S, Hanson M, Pavirani A, Rodeck C, Coutelle C (1999) High level expression of β-galactosidase and factor IX transgenes in fetal and neonatal sheep after ultrasound-guided transcutaneous adenovirus vector administration into the umbilical vein. Gene Ther 6:1239–1248

Türkay A, Saunders TL, Kurachi K (1999) Intrauterine gene transfer: gestational stage-specific gene delivery in mice. Gene Ther 6:1685–1794

Vincent MC, Trapnell B, Baughman RP, Wert SE, Whitsett JA, Iwamoto HS (1995) Adenovirus-mediated gene transfer to the respiratory tract of fetal sheep in utero. Hum Gene Ther 6:1019–1028

Wang G, Williamson R, Mueller G, Thomas P, Davidson BL, McCray PB (1998) Utrasound-guided gene delivery to hepatocytes in utero. Fetal Diagn Ther 13:197–205

Wang L, Takabe K, Bidlingmaier SM, Ill CR, Verma IM (1999) Sustained correction of bleeding disorder in hemophilia B mice by gene therapy. Proc Natl Acad Sci USA 96:154–158

Woo Y, Raju G, Swain J, Richmond M, Gardner T, Balice-Gordon RJ (1997) In utero cardiac gene transfer via intraplacental delivery of recombinant adenovirus. Circulation 96:3561–3569

Yang E, Cass DL, Sylvester KG, Wilson JM, Adzick NS (1999a) Fetal gene therapy: efficacy, toxicity, and immunologic effects of early gestation recombinant adenovirus. J Pediatr Surg 34:235–241

Yang E, Kim HB, Shaaban AF, Milner R, Adzick NS, Flake AW (1999b) Persistent postnatal transgene expression in both muscle and liver after fetal injection of recombinant adenovirus. Pediatr Surg 34:766–772; discussion 772–773

8 Prenatal Transplantation
of Hematopoietic Stem Cells: Overview

D.V. Surbek, W. Holzgreve

8.1 Introduction

In the last 25 years, extensive progress has been made in the prenatal diagnosis of congenital diseases using non-invasive and invasive techniques. In contrast to prenatal diagnosis, prenatal therapy has shown limited success so far. Therefore, if a severe congenital disease is diagnosed early in gestation, many parents choose to terminate pregnancy. Prenatal transfer of hematopoietic stem cells (HSC) is a promising approach to successfully treat fetuses affected by hematologic, immunologic, or metabolic diseases. Recently, clinical success has been achieved (Flake et al. 1996). The success, however, is limited to diseases where severe immunodeficiency is present in the fetus. This article aims to provide an overview of the current experience in animals and humans and to identify strategies to overcome the current obstacles.

8.2 Rationale for In Utero Transplantation

Postnatal HSC transplantation is a successful treatment for many genetic diseases of the hematopoietic and immunologic system as well as for certain storage diseases. However, lack of a human leukocyte antigen (HLA)-compatible donor, graft rejection or graft versus host disease (GvHD), and morbidity associated with previous myeloablation and post-transplant immunosuppression are major disadvantages. Furthermore, preexisting organ damage acquired in utero may add to procedure-related toxicity.

The physiology of the development of the hematopoietic and immunologic system in human fetuses offers a theoretical opportunity to circumvent these problems (Surbek et al. 1999). Early in gestation, until the end of the first trimester, the fetus is immunologically naive, theoretically obviating the need for HLA-matching and immunosuppression of transplanted cells and possibly inducing donor-specific tolerance, which is regulated by the fetal thymic microenvironment. Hematopoietic ontogenesis is characterized by a chronological sequence of site change from the yolk sac and the aorta-gonadal mesonephron region to the fetal liver early in gestation and thereafter to the fetal bone marrow, which is still relatively empty before the second trimester, a fact which obviates the need for marrow ablation prior to stem cell transplantation. Other advantages may include the sterile, protecting environment in utero and the prevention of prenatally developing irreversible organ damage. Probably the best evidence for the feasibility of prenatal HSC transplantation in immunocompetent individuals comes from reports of naturally occurring chimerism in dizygote twins sharing transplacental circulation in utero, which has been described in animals and more recently in humans. Successful in utero transplantation might therefore not be restricted to diseases leading to severe immunodeficiency. Therefore, every disease which is treatable by postnatal bone marrow transplantation and which can be diagnosed prenatally could be a candidate disease. Table 1 contains a partial list of diseases which might be potentially amenable to in utero HSC transplantation.

Table 1. Genetic diseases potentially amenable to in utero hematopoietic stem cell transplantation

Hemoglobinopathies
α-Thalassemia major
β-Thalassemia major
Sickle cell anemia
Immunodeficiency disorders
Severe combined immunodeficiency syndrome
(SCID; X-linked, aminodeaminase-deficiency or other defects)
Wiskott-Aldrich syndrome
Chronic granulomatous disease
Others (Fanconi anemia)
Metabolic diseases
M. Hurler
Maroteaux-Lamy syndrome
Sly's syndrome
M. Gaucher
M. Krabbe
Metachromatic leukodystrophy
M. Niemann-Pick
Others

8.3 Animal Models

Based on the concept of acquired neonatal tolerance to foreign antigens, initial experiments have been performed in the severely anemic mouse model using intraplacental injection of HSC, showing the potential of prenatal transplantation in hematopoiesis-deficient animals. Later, other mouse models with impaired immune system [e.g., non-obese diabetic (NOD)/severe combined immunodeficiency syndrome (SCID) mouse] have been successfully used as a model for prenatal allogeneic HSC transplantation. In the non-deficient fetal sheep model, long-term stable engraftment has also been achieved using allogeneic as well as xenogeneic (human) HSC transplantation. However, it soon became evident that in other non-deficient animal models, e.g., monkeys, goats, or mice, the levels of chimerism – if engraftment could be achieved at all –

remain very low (%). Furthermore, even in a sheep model with significant HSC engraftment level, a beneficial effect on prenatal central
nervous system (CNS) disease could not be shown, which may limit the
usefulness of prenatal transplantation for metabolic diseases affecting
the CNS. Nevertheless, many of these models have proven to be useful
to study the mechanism and kinetics of engraftment of in utero transplanted cells and to test strategies to overcome current obstacles.

8.4 Clinical Experience

To date, up to about 30 cases of in utero HSC transplantation have been
reported, one-third being performed for immunodeficient disorders including SCID or chronic granulomatous disease (Flake and Zanjani
1999). The remaining cases included fetuses affected by Rhesus isoim-
munization, hemoglobinopathies, or storage diseases (e.g., Hurler syndrome). Different sources of donor stem cells were used (maternal,
paternal or sibling bone marrow and fetal liver cells) with or without T
cell-depletion/CD34$^+$ cell enrichment. The gestational age of the recipients ranged from 11 to over 30 weeks, and in some cases, multiple
transplants were performed. Significant engraftment has been achieved
in fetuses affected by a severe deficiency of the immune system. Two
recent reports in fetuses with X-linked SCID showed successful treatment; both have persistent split chimerism (donor T and NK cells, host
B cells) after birth. In contrast, in fetuses without immunodeficiency, no
clinically significant, stable engraftment could be achieved, although
some microchimerism and donor-specific tolerance could be shown in
some cases. In one case of prenatal HSC transplantation in a fetus with
globoid cell leukodystrophy, fetal death occurred 6 weeks after the
procedure. Autopsy findings revealed donor cells in most fetal organs,
consistent with engraftment and/or GvHD.

   In summary, stable engraftment in non-myeloablated immunologically competent host fetuses has not been achieved so far. This might be
due to the lack of competitive advantage of donor hematopoietic cells
over host cells in bone marrow niches, and/or to graft rejection by the
host.

8.5 Recent Strategies

Current strategies to improve donor cell engraftment are based on improved understanding of the biology of in utero transplantation obtained from experiments in animals and experience in humans.

The *timing of transplantation* has been recognized to be a most important issue in non-immunodeficient fetuses. The gestational age where the fetus is believed to be immunoincompetent, not rejecting foreign antigens, is believed to be before 14 weeks; thereafter, mature immunocompetent T lymphocytes are present in peripheral fetal blood.

The *source of stem cells* might be of some importance. Some evidence suggests that fetal stem cells from fetal liver or fetal bone marrow may be superior to adult sources like bone marrow or peripheral blood, especially in selected diseases such as hemoglobinopathies. There are, however, major drawbacks involved in the use of fetal cells, including ethical aspects, the risk of infection in the recipient, and the limited amount of cells available from the same donor. Cord blood is another important source of fetal cells which might have favorable characteristics (Surbek et al. 1998); it is increasingly used in the postnatal setting. Clinical success has been achieved with adult sources. If adult bone marrow or cord blood is used, $CD34^+$ cell enrichment and T cell depletion are necessary.

Another potentially important issue is the *dose of HSC* used. In the fetal sheep model, dose-dependent increase of engraftment level seemed to reach a plateau above a certain dose. However, evidence from postnatal transplantation experiments in mice suggest that large donor cell doses might displace host cells from niches in the bone marrow, provided that they are equally competitive. This aim might be achieved by the use of repetitive transplantation, which have been already successfully used in human fetuses with X-SCID. The *route of administration* (intraperitoneal vs intravenous) might further influence the HSC dose in the target organ, the fetal bone marrow; however, safety aspects must be taken into consideration, especially if multiple procedures are performed.

Graft modification by co-transplantation of donor-specific stromal cells is probably the most promising strategy to date. It has been shown not only to persistently increase donor cell level, but also early in utero donor cell expression in fetal blood in the sheep in utero transplantation model (Almeida-Porada et al. 1999). Especially in diseases which lead to early organ damage, the fetus might benefit from early donor cell "expression" in the periphery. The addition of specific hematopoietic growth factors or blocking antibodies to support engraftment of donor cells might be an option supported by recent experimental evidence (Zanjani et al. 1999).

Hematopoietic *microchimerism* has recently been identified in recipients of solid-organ transplants and is thought to be essential for maintenance of immunological unresponsiveness to donor organs. The concept of *tolerance induction* and mixed chimerism is currently being investigated clinically using simultaneous transplantation of bone marrow-derived cells and solid organ transplantation from the same donor to induce donor-specific tolerance (Elwood et al. 1998). After in utero stem cell transplantation, persistent microchimerism has been achieved in animal models as well as in humans (Thilaganathan and Nicolaides 1993). Donor-specific tolerance was shown in vitro in humans and in vivo in mice using later skin graft or in primates using later kidney graft from the same donor. As data from other preclinical studies suggest, prenatally induced tolerance can be used for *postnatal "boost"-transplantation* of stem cells from the same donor without or with minimal conditioning. In diseases which do not induce severe immunodeficiency, this strategy might be a sensible and potentially successful approach. The clinical value of this procedure remains to be determined.

A completely different approach to overcome important hurdles in prenatal allogeneic stem cell transplantation might be *fetal gene therapy*. Two different strategies for gene transfer are under investigation. One is to obtain fetal cells for in vitro transduction and autologous transplantation, as has been done with partial success with cord blood in neonates with aminodeaminase (ADA) deficiency. The other possibility is to directly transfer the gene-containing vector to the fetus. Although fetal gene therapy might be the clue for successful prenatal treatment of many diseases, major hurdles have to be overcome, predominantly efficient HSC transduction and long-term gene expression. Especially if the vecto-gene construct is directly administered to the fetus, germ line transduction or maternal cell transduction can theoretically occur and must be excluded. Prior to the use of fetal gene therapy in humans, these issues must at least partially be resolved, and concomitant ethical questions must be addressed (Zanjani and Anderson 1999).

8.6 Comment

Prenatal HSC transplantation holds much promise for the treatment of a variety of severe congenital diseases. This promise has been fulfilled so far only in immunological diseases, despite the "preimmune" fetus being theoretically amenable for stem cell transplantation. Better understanding of the physiology and disease-specific impairment of fetal immunological development and of the molecular mechanism of microenvironment–stem cell interaction during the hematopoietic homing process in the fetal bone marrow after HSC transplantation, together with clinical experience will pave the way to successful treatment. Ethical issues need to be included in clinical protocols for in utero transplantation, currently being established in many centers. They will form a basis for the appropriate selection of disease categories and individual patients and parents.

References

Almeida-Porada G, Flake AW, Glimp HA, Zanjani ED (1999) Cotransplantation of stroma results in enhancement and early expression of donor hematopoietic stem cells in utero. Exp Hematol 27:1569–1575

Elwood ET, Larsen CP, Maurer DH et al (1998) Microchimerism and rejection in clinical transplantation. Lancet 349:1358–1360

Flake AW, Zanjani ED (1999) In utero hematopoietic stem cell transplantation: ontogenetic opportunities and biologic barriers. Blood 94:2179–2191

Flake AW, Roncarolo MG, Puck JM et al (1996) Treatment of X-linked severe combined immunodeficiency by in utero transplantation of paternal bone marrow. N Engl J Med 335:1806–1810

Surbek DV, Holzgreve W, Jansen W et al (1998) Quantitative immunophenotypic characterization, cryopreservation, and enrichment of second and third trimester human fetal cord blood hematopoietic stem/progenitor cells. Am J Obstet Gynecol 179:1228–1233

Surbek DV, Gratwohl A, Holzgreve W (1999) In utero hematopoietic stem cell transfer: current status and future strategies. Eur J Obstet Gynecol Reprod Biol 85:109–115

Thilaganathan B, Nicolaides KH (1993) Intrauterine bone-marrow transplantation at 12 weeks' gestation. Lancet 342:243

Zanjani ED, Anderson WF (1999) Prospects for in utero human gene therapy. Science 285:2084–2088

Zanjani ED, Flake AW, Almeida-Porada G, Tran N, Papayannopoulou T (1999) Homing of human cells in fetal sheep model: modulation by antibodies activating or inhibiting very late activation antigen-4-dependent function. Blood 94:2515–2522

9 Fetal Hematopoietic Stem Cells: In Vitro Expansion and Transduction Using Lentiviral Vectors

A. Luther-Wyrsch, C. Nissen, D.V. Surbek, W. Holzgreve,
E. Costello, M. Thali, E. Buetti, A. Wodnar-Filipowicz

9.1 Introduction

Umbilical cord blood (CB), remaining in the placenta at delivery, is rich in hematopoietic stem and progenitor cells. Compared with adult peripheral blood, the content of $CD34^+$ and $CD34^+CD38^-$ cells in CB is approximately tenfold higher and thus comparable to adult bone marrow (Broxmeyer et al. 1989, 1990; Hao et al. 1995). Moreover, CB progenitors have high plating efficiency in clonogenic assays, they respond rapidly to cytokine stimulation in vitro and generate progeny comparable to that derived from bone marrow precursors (Emerson et al. 1985; Cardoso et al. 1993; Lansdorp et al. 1993). Due to these characteristics, CB has been recognized as an attractive alternative source of hematopoietic stem cells for transplantation (Cairo and Wagner 1997). Several hundreds of patients with hematological malignancies and genetic diseases affecting the hematopoietic system have been treated with

CB from related as well as from unrelated donors (Gluckman et al. 1989, 1997; Rubinstein et al. 1998). Clinical results have shown that CB transplants engraft and sustain hematopoietic function. Survival rates are comparable to those after bone marrow transplantation. The main advantage of CB over adult hematopoietic tissue is immunologic immaturity of accessory cells in the graft, including a lower expression level of T cell-derived growth factors (Ehlers and Smith 1991; Harris et al. 1992; Sautois et al. 1997). In clinical practice, this results in reduced incidence and severity of graft-versus-host disease, and may allow a higher degree of human leukocyte antigen (HLA) disparity between donor and recipient (Rubinstein et al. 1998). While CB transplantation has proven feasible in pediatric patients, the number of stem cells in a CB graft may be insufficient to reconstitute an adult recipient. Therefore, efforts to expand pluripotent hematopoietic CB cells in vitro are under way. The most efficient amplification of hematopoietic cells from human CB has recently been described in cultures supplemented with thrombopoietin, and flt-3 ligand (Moore and Hoskins 1994; Piacibello et al. 1997).

CB-derived stem cells are an attractive target for somatic gene therapy of inborn defects of the lymphoid and the hematopoietic system. Engraftment of genetically modified CD34+ cells has been achieved with CB of neonates with adenosine deaminase deficiency (Kohn et al. 1995; Bordignon et al. 1995). Advances in understanding the molecular background of hereditary diseases and progress in prenatal diagnostics have opened a therapeutic concept of in utero treatment of genetic diseases identified during pregnancy (Flake and Zanjani 1997). Feasibility of in utero transplantation has been documented by long-term multilineage chimerism in sheep, mice and monkeys transplanted with human tissue early during gestation (Flake et al. 1986; Fleischman and Mintz 1979; Harrison et al. 1989; Zanjani et al. 1992). Recently, first clinical reports described the successful treatment of immunodeficiency syndromes in human fetal recipients transplanted with parental bone marrow (Flake et al. 1996; Wengler et al. 1996) or allogeneic fetal liver cells (Touraine et al. 1989, 1992). A therapeutic option of in utero therapy could also be envisaged with autologous fetal stem cells if enough circulating progenitors could be acquired by cordocentesis during pregnancy, genetically modified ex vivo, and transplanted back to the affected fetus. This strategy would avoid limitations and risks associated with immunological barriers of allogeneic grafts. So far, progress in human gene therapy has been hampered by inefficiency of gene transfer to immature hematopoietic progenitors with long-term repopulating capacity after transplantation. Recently, new generations of the replication-deficient lentiviral vectors have been shown useful for ex vivo gene delivery to non-dividing cells (Naldini et al. 1996; Uchida et al. 1998), thus opening a perspective for targeting quiescent hematopoietic progenitors.

While hematopoietic properties of neonatal CB from full-term pregnancies have been well characterized, less is known about CB from early gestational ages. Previous studies revealed the presence of hematopoietic cells in the fetal circulation: it has been demonstrated that the content of CD34$^+$ progenitors in CB is higher during fetal life than at birth (Thilaganathan et al. 1994; Shields and Andrews 1998; Surbek et al. 1998), and that preterm CB is rich in colony-forming precursors (Clapp et al. 1989; Jones et al. 1994; Migliaccio et al. 1996). In our recent work, we analyzed the primitive hematopoietic progenitors, defined phenotypically as CD34$^+$CD38$^-$ cells and functionally as long-term culture-initiating cells (LTC-IC). Using different combinations of recombinant growth factors, we have also established culture conditions for short- and long-term expansion of circulating progenitors from preterm CB. The results demonstrated that CB from fetuses at early gestational age contains a significantly higher number of both committed as well as early progenitor cells than term CB, with profound proliferation capacity in vitro (Wyrsch et al. 1999). In this work, we used lentiviral vectors for transfer of the enhanced green fluorescence protein (GFP) gene to fetal progenitors. We show that the efficiency of transduction of hematopoietic cells from preterm and term CB is comparable. Furthermore, transduced cells can be extensively amplified in vitro. Owing to these properties, preterm fetal CB may be a potential source of progenitors for in utero treatment of disorders amenable to transplantation of genetically corrected stem cells.

9.2 Material and Methods

9.2.1 Study Population

The samples from term healthy newborns (n=20; weeks ≥35) included
uneventful vaginal births and cesarean sections. Preterm CB included
samples from weeks 13 and 14, from second trimester (n=18;
weeks 16–28), and early third trimester (n=20; weeks 29–34) of pregnancy. Premature deliveries followed idiopathic preterm labor or pre-
labor rupture of membranes, as well as spontaneous and elective abortions in the second trimester. Exclusion criteria were clinically overt
chorioamnionitis and preeclampsia. There has been no evidence of
hematopoietic abnormalities of the fetuses. Informed consent was obtained from the mothers prior to delivery, and the Ethical Committee of
the University Hospitals in Basel, Switzerland approved the investigations.

9.2.2 Cord Blood Cells

CB was harvested aseptically by umbilical vein puncture, and up to
10 ml was collected in heparin-containing tubes. The delay between
collection of CB samples and their subsequent processing did not ex-
ceed 12 h. Flow cytometry was performed with 250 µl of heparinized
full blood. For cell cultures, cord blood mononuclear cells (CBMNC)
were isolated by centrifugation on Histopaque (density <1.077 g/ml;
Sigma Diagnostics, St. Louis, Mont., USA), and used either freshly to
initiate methylcellulose cultures (see below), or were cryopreserved in
liquid nitrogen in a freezing mixture containing Iscove's modified Dul-
becco's medium (IMDM), 20% fetal calf serum (FCS; both Gibco,
Gaitherburg, Md., USA) and 10% DMSO and used for long-term cultures or for transduction with lentiviral vectors. Prior to the experiment,
cells were thawed and allowed to adhere overnight in culture dishes
containing IMDM-25% FCS; nonadherent cells were collected to initiate the cultures (see below).

9.2.3 Flow Cytometry (FACS) Analysis

Aliquots of 50 µl of heparinized CB were stained with anti CD34-phycoerythrin (PE; Becton Dickinson, San Jose, Calif., USA) and anti CD38-fluoroscein isothiocyanate (FITC; Immunotech, Marseille, France) or with the corresponding isotype control antibodies (Becton Dickinson) at concentrations recommended by the manufacturers. After staining, the samples were treated with lysing buffer (Ortho, Neckargemünd, Germany) to lyse red blood cells. Dual color analysis was performed on a FACScan (Becton Dickinson) acquiring 10–50,000 events. Lymphocyte gates were set according to forward and sideward light scatter; dead cells were stained with propidium iodide and excluded from the analysis. Analysis was performed with CellQuest software (Becton Dickinson).

9.2.4 Growth Factors

The following recombinant human growth factors were used: stem cell factor (SCF; AMGEN, Thousand Oaks, Calif., USA), megakaryocyte growth and development factor (PEG-rHuMGDF, a truncated Mpl ligand, AMGEN), flt-3 ligand (FL; Immunex, Seattle, Wash., USA), Interleukin (IL)-3, -6, granulocyte-macrophage colony-stimulating factor (GM-CSF; all from Novartis, Basel, Switzerland), granulocyte colony-stimulating factor (G-CSF; Rhone-Poulenc, Antony, France), and erythropoietin (Epo; Böhringer Mannheim, Mannheim, Germany). Growth factors were used at a concentration of 100 ng/ml, except for MGDF at 50 ng/ml and Epo at 36 U/ml.

9.2.5 Cell Culture Assays

Colony forming Unit Assay. CBMC from term and preterm samples (5×10^4 and 2.5×10^4, respectively) were plated into 1% methylcellulose cultures (Wodnar-Filipowicz et al. 1992) supplemented with the growth factors indicated in the figure legends. Secondary methylcellulose cultures, performed with cells harvested from liquid cultures (see below), were supplemented with Epo, G-CSF, GM-CSF, IL-3, and SCF. Hematopoietic colonies derived from granulocyte (G)-colony forming units (CFU), macrophage (M-CFU), erythroid (BFU-E), granulocyte-macrophage (GM-CFU), granulocyte-macrophage-erythroid-mega-karyocyte (GEMM-CFU), and megakaryocyte (Mk-CFU) precursors were counted after 14 days of culture. The identity of Mk-CFU was confirmed by cell morphology after differential staining (Dade Diff-Quik, Baxter Diagnostics, Düdingen, Switzerland) of individually picked colonies.

Long-Term Culture-Initiating Cell Assay. Determination of long-term culture-initiating cell (LTC-IC) content was performed according to Sutherland et al. (1990). The murine fibroblast cell line M210B4 (American Type Culture Collection) was used as a feeder layer. Before initiation of the co-culture with human cells, M210B4 cells were trypsinized, irradiated with 80 Gy, seeded into 96-well plates and al-lowed to readhere over night. Term or preterm CBMNC (6×10^3 and 3×10^3, respectively) were seeded in quadruplicate wells over the feeders in 200 µl of IMDM containing 12.5% horse serum, 12.5% FCS, 0.016 mm folic acid, 0.16 mm i-inositol (all from Gibco) and 10^{-6} m hydrocortisone (Sigma Diagnostics). If limiting dilution analysis was carried out, a minimum of 12 replicate wells per cell concentration ($50-3\times10^3$ CBMNC/well) were initiated. The cultures were maintained at 33°C for 5 weeks with weekly half-medium changes. At the end of the culture period, nonadherent cells were combined with the corresponding trypsinized adherent cells, washed and assayed in secondary methylcel-lulose cultures for CFUs as described above.

Long-Term Liquid Cultures. Cultures were carried out according to Piacibello et al. (1997). Briefly, 1×10^5 term or preterm CBMNC were seeded into 1 ml of IMDM containing 10% FCS, 380 mg/ml iron satu-rated human transferrin, 1% bovine serum albumin (BSA; Behring, Marburg, Germany). Cultures were supplemented with MGDF and FL or SCF. Weekly half-medium changes were performed. At 1, 3, 6, and 10 weeks, harvested cells were counted, and aliquots were plated into secondary methylcellulose cultures to determine the content of CFUs. In addition, at 1, 3 and 6 weeks, cells were seeded over M210B4 feeder layers, and the content of LTC-IC was determined.

9.2.6 Transduction with Lentiviral Vectors

Lentiviral vectors were produced by co-transfection of 293 T cells with three plasmids, as described in Costello et al. (1999), and concentrated by ultracentrifugation to 5×10^7 IU/ml. Prior to transduction, 1×10^5 CD34$^+$ CB cells were precultured for 3 days with IL-3, -6, SCF, and FL. For transduction, cells were resuspended in 50 µl IMDM with 5 µg/ml protamine sulfate (Sigma, St. Louis, USA) in V-bottom tubes previously coated with 20 µg/cm^2 fibronectin (Retronectin, TaKaRa, Shiga, Japan) or BSA. Lentivirus-containing supernatant (30 µl) was added, and cells were centrifuged at 3000 rpm for 3 h (spinoculation). Cells were then resuspended in 1 ml of IMDM containing 20% FCS and growth factors as above, and incubated without washing for 24 h at 37°C. The next day, cells were washed twice with IMDM supplemented with 2% FCS. In some experiments, transduction was repeated. Samples were spinoculated with fresh lentiviral supernatant, incubated and washed as described above. Transduced CB cells were counted and used for CFU-, LTC-IC- and liquid cultures. GFP-expression was analyzed by flow cytometry starting from day 3 after transduction.

9.2.7 Statistical Analysis

Growth of term and preterm CB was compared by the Student's unpaired t-test, and a linear correlation test was performed using StatView software (Abacus).

9.3 Results

9.3.1 Content of CD34$^+$ and CD34$^+$CD38$^-$ Progenitors

The frequency of CD34$^+$ cells and their CD34$^+$CD38$^-$ subset was determined in CB from fetuses at different gestational ages, ranging from 13th week to term (35th week) pregnancies. We were unable to detect hematopoietic progenitors in samples derived from late first trimester of pregnancy (weeks 13 and 14). In contrast, high content of progenitors was measured in samples from the second trimester, starting from

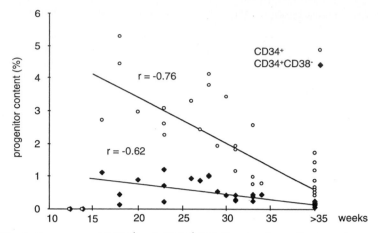

Fig. 1. Frequency of CD34$^+$ and CD34$^+$CD38$^-$ progenitor cells along the progression of pregnancy

week 16 (Fig. 1). The percentage of CD34$^+$ and CD34$^+$CD38$^-$ cells decreased steadily, in an inverse correlation to gestational age (r=–0.76 and –0.62, respectively). On average, the content of CD34$^+$ cells in preterm CB was 2.51±0.28%, which was significantly higher in both second and early third trimester than in CB from newborns delivered at term (0.88±0.17%; $P<0.001$). Preterm CB was also richer in the most primitive CD34$^+$CD38$^-$ progenitors (0.56±0.08 vs 0.13±0.02%; $P<0.002$) (Wyrsch et al. 1999).

9.3.2 Content of CFU and LTC-IC Progenitors

To define the frequency of committed CFU progenitors, we cultured CBMNC of various gestational ages in methylcellulose supplemented with a combination of growth factors containing Epo, IL-3, G-CSF, GM-CSF and SCF (Wodnar-Filipowicz et al. 1992) (Table 1). Colony numbers were approximately twofold higher in preterm than term CB (240±27 vs 142±12/10^5 CBMNC; $P<0.05$), with a prevalence of erythroid and mixed colonies (BFU-E and GEMM-CFU). The amount of CFUs correlated with the frequency of CD34$^+$ cells in CB (r=0.73;

Table 1. Frequency of hematopoietic colony-forming cells in term and preterm CB

CB	CFU[a]	LTC-IC[b]
Term	142±12	2.6±1.2
Preterm	240±27	6.7±2.9

Mean±SEM is indicated. Statistical significance of the differences between CFU and LTC-IC content in term and preterm CB, $P<0.05$.
[a] Frequency of CFUs per 1×10^5 CBMNC from term and preterm CB. CBMNC were cultured in methylcellulose in the presence of Epo, IL-3, G-CSF, GM-CSF and SCF.
[b] Frequency of LTC-ICs per 1×10^5 CBMNC from term and preterm CB. Results are based on limiting dilution analysis.

$P<0.0004$). For detection of more primitive LTC-IC precursors, stroma-supported long-term cultures were established. LTC-IC-derived colonies were detected in every preterm CB sample (Table 1). Their frequency was $6.7\pm2.9/10^5$ CBMNC, which was higher than in term CB (2.6 ± 1.2; $P<0.05$).

9.3.3 Comparison of Hematopoietic Progenitor Cells from Various Sources

Table 2 summarizes published results on the content and growth properties of progenitors from various hematopoietic tissues. The frequency of CD34+ cells in preterm CB, although highly variable, is higher than in traditional sources of transplantable stem cells such as "mobilized" peripheral blood or bone marrow. Significantly higher is also the content of CD34+CD38− cells, as well as of progenitors scored in the functional hematopoietic in vitro assays, CFU and LTC-IC. Preterm CB has not yet been examined in terms of the content of SCID repopulating cells (SCR), which represent primitive human hematopoietic cells capable of in vivo repopulating the bone marrow of sublethally irradiated mice with nonobese diabetic/severe combined immunodeficiency (NOD/SCID) syndrome (Larochelle et al. 1996).

Table 2. Comparison of the frequency and growth properties of hematopoietic progenitor cells from different tissues

Source	CD34+ (% MNC)	CD34+CD38– (% MNC)	CFU /10⁶ MNC	LTC-IC /10⁶ MNC	SRC /10⁶ MNC	References
PB	0.1	<0.1	40	0.5	<	Civin et al. 1987
PB "mobilized"	0.5–1.0	0.1	820	60	0.2	Wang et al. 1997, Hogge et al. 1996
BM	1.0–4.0	0.1	1400	30	0.3	Civin et al. 1987, Wang et al. 1997
CB term	0.5–2.0	0.1	1300	70	1	Hao Q-L et al. 1995, Wang et al. 1997
CB preterm	1.0–15.0	0.6	2300	150	?	Wyrsch et al. 1999

PB, peripheral blood; *BM*, bone marrow; *MNC*, mononuclear cells; *CB*, cord blood.

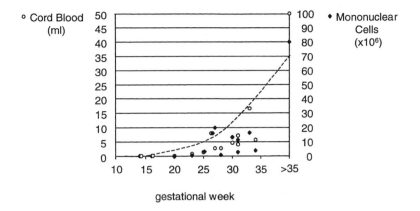

Fig. 2. Yield of cord blood and of mononuclear cells along the progression of pregnancy. Increasing weight of the fetus is marked with the *dotted line*

9.3.4 Retrieval of Hematopoietic Progenitor Cells from Preterm CB

We assessed the yield of hematopoietic progenitors from fetal compared with term CB, taking into account the total volume, the recovery of mononuclear cells, and the number of progenitors measured in samples from various gestational ages. The volume of CB collected from the umbilical cord and the total content of nucleated cells paralleled the increasing weight of the fetus (Fig. 2). On average, the CB volume and the cell content in second and third trimester samples were approximately 25- and 8-fold lower than in CB from term pregnancies (Ta-

Table 3. Total retrieval of hematopoietic progenitors from preterm and term CB

	Volume (ml)	Mononuclear cells ($\times 10^6$)	CD34$^+$ cells ($\times 10^5$)	CFUs
2nd trimester CB (until week 28)	2.3	3.3	1.0	8000
Early 3rd trimester CB (week 29–35)	8.0	10.3	2.1	21100
Term CB (>week 35)	50.0	80.0	7.4	106,000

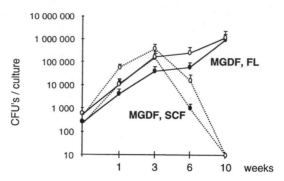

Fig. 3. Long-term expansion of CFU progenitors in cultures containing MGDF and FL or MGDF and SCF, as indicated. (o) preterm CB; (●) term CB

ble 3). Therefore, despite the high frequency of hematopoietic progenitors in fetal CB, the total yield of CD34$^+$ cells and CFU-derived colonies was significantly below the yield of progenitors from term CB (Table 3).

9.3.5 Long-Term Expansion of Clonogenic Progenitors

Recently, a dramatic effect of TPO in combination with FL on the expansion of CB precursors ex vivo has been reported by Piacibello et al. (1997). We were able to reproduce this effect with preterm CB: in the presence of FL and MGDF, representing a functional equivalent of TPO (Hunt et al. 1995), inexhaustible production of CFUs was observed in stroma-free liquid cultures monitored for 10 weeks (Fig. 3). An approximately 1000-fold expansion of CFUs was achieved in preterm as well as term CB. At week 10, multipotential progenitors (GEMM-CFU) were still prevailing in both preterm and term CB (Wyrsch et al. 1999). The effect of MGDF and FL on the expansion of LTC-IC was monitored until week 6; an approximately 30-fold amplification was observed in preterm CB cultures (Fig. 4). When MGDF was used together with SCF, clonogenic progenitors started to decline at week 3 and were exhausted after 10 weeks (Fig. 3). These results indicate that preterm CB progenitors can be not only maintained but also extensively amplified in cultures in the presence of appropriate combinations of growth factors.

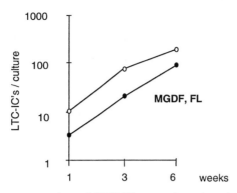

Fig. 4. Long-term expansion of LTC-IC progenitors in cultures containing MGDF and FL. (o) preterm CB; (●) term CB

9.3.6 Transduction of CB Cells with Lentiviral Vectors

We used preterm CB cells as targets for transfer of the enhanced GFP gene by lentiviral vectors. Preparation of the vectors has been described elsewhere (Costello et al. 1999). Various gene transfer protocols were tested for transfection of purified CD34+ progenitors. Efficiency and longevity of GFP-expression were determined by flow cytometry and fluorescence microscopy of colonies in CFU assays. Best results were obtained by exposing cells to viral supernatant supplemented with protamine sulfate in 3-h spinoculation at 3000 rpm on 2 subsequent days (Fig. 5). In liquid cultures, 20–30% of CD34+ cells was expressing GFP at 2–3 weeks after transfection, and this expression persisted for at least 5 weeks. The frequency of transduced progenitors from preterm and term CB was comparable. Likewise, the frequency of GFP-expressing CFU-derived colonies from preterm and term CB was similar (Table 4). About 25% of colonies obtained prior to expansion in MGDF/FL-containing cultures were GFP+. This high frequency of transduced cells was maintained in cultures examined at 3 weeks. Simultaneously, a significant amplification of clonogenic progenitors was observed, in agreement with the results shown in Fig. 3.

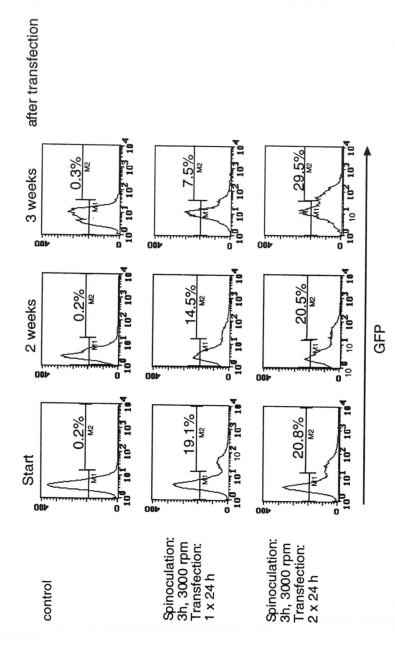

Fig. 5. Flow cytometry analysis of CB progenitor cells transduced with GPF gene using lentiviral vectors. Transduction conditions and the time of analysis are indicated

Table 4. GFP-expression in lentivirally transduced clonogenic CB progenitors in response to MGDF and FL

		Term CB (n=2)	Preterm CB (n=4)
Start	CFUs/10^3 CD34$^+$ cells	176±21	153±6
	GFP$^+$ CFUs	41±7	35±6
	%	23±1	23±5
3 weeks	CFUs/10^3 CD34$^+$ cells	10374±375	10100±2750
	GFP$^+$ CFUs	2475±175	4020±1150
	%	24±1	42±7

Transduced cells were cultivated in liquid cultures containing MGDF and FL, and assayed in secondary methylcellulose cultures at week 3. The content of GFP$^+$CFUs was determined by fluorescence microscopy.

9.4 Discussion

The first goal of this study was the phenotypic and functional characterization of committed and primitive hematopoietic CB progenitors from immature infants at weeks 13–34 of gestation, as compared with CB from newborns delivered at term. The frequency of circulating progenitor cells was the highest during the second trimester of pregnancy and declined linearly with progressing gestation. The content of CD34$^+$ cells decreased from an average of 3.11% to 0.88%, CD34$^+$CD38$^-$ from 0.69% to 0.13%, CFUs committed to granulocyte, macrophage and erythroid lineages from 230.2 to 133.2/10^5 CBMNC, and clonogenic LTC-ICs from 6.7 to 2.6/10^5 CBMNC. The gestation time-dependent changes in the content of circulating progenitors are in accordance with the sequential shifts of hematopoietic sites during ontogeny: At 6 weeks of gestation, hematopoiesis switches from yolk sac and aorta-gonad-mesonephros region to the fetal liver, which remains the major organ of blood cell production throughout pregnancy. During the second and third trimester, circulating stem cells start colonizing bone cavities and gradually establish adult hematopoiesis in the bone marrow (reviewed in Tavassoli 1991 and Péault 1996). According to our results, the highest concentration of progenitors is observed during transition from hepatic to bone marrow hematopoiesis, and the decline occurs with the termination of this process at birth.

The frequency of progenitors in fetal CB exceeded that reported in bone marrow and peripheral blood after "mobilization" (Civin et al. 1987; Udomsakdi et al. 1992; Hogge et al. 1996; Wang et al. 1997). However, the retrieval of CB from premature deliveries was low in terms of blood volume and mononuclear cell harvest. Consequently, despite the high frequency of hematopoietic progenitors in fetal CB, the total yield of CD34+ cells and clonogenic progenitors was significantly below the yield observed with term CB. To overcome the limitations of the low cell numbers, we investigated the ability of preterm CB progenitors to undergo amplification ex vivo. We demonstrate that hematopoietic progenitors from preterm fetal CB can be expanded in vitro in cultures supplemented with early-acting hematopoietic growth factors. In short-term liquid cultures containing Epo, IL-1, -3, and -6, or G- and GM-CSF together with SCF or FL, expansion of CFUs was six- to eightfold at week 1 (Wyrsch et al. 1999). In long-term cultures containing MGDF and FL, an approximately 1000-fold expansion of multilineage progenitors was observed until week 10. The expansion of preterm progenitors was equally efficient as of term progenitors. Our results show that stem cells from CB of preterm deliveries preserved full proliferative capacity in vitro, although the fetuses might have suffered from unfavorable conditions before or during birth, either from a primary disease leading to premature termination of pregnancy, or from oxygen deprivation during induced abortion.

The second goal of this study was to examine the efficiency of retrovirally mediated gene transfer to fetal CB progenitors. Recently, new generations of retroviral vectors, including the replication-deficient lentiviral vectors, have been shown useful for ex vivo gene delivery to nondividing cells (Naldini et al. 1996; Uchida et al. 1998). CB progenitors are among the target cells in which sustained expression of transferred genes has been achieved (Conneally et al. 1998; van Hennik et al. 1998). This includes the highly efficient lentiviral transduction of CB-derived SRC progenitors capable of long-term repopulation of NOD/SCID mice (Miyoshi et al. 1999). We demonstrate that fetal CB progenitors can be efficiently transduced with lentiviral vectors carrying the GFP gene. GFP expression was found in approximately 25% of transduced CD34+ cells, as well as CFU and LTC-IC-derived colonies. These high levels of GFP+ cells were maintained during long-term in vitro expansion in MGDF/FL-containing medium. The frequency of transduced cells from preterm and term CB was comparable.

Studies on proliferation, expansion and transduction properties of CB progenitors from early gestational ages are relevant with regard to their potential use as autologous grafts in utero. Prenatal somatic gene therapy with autologous genetically corrected hematopoietic stem cells may have distinct advantages over postnatal therapy in genetic diseases such as thalassemias, immunodeficiency syndromes, or selected metabolic disorders. In these diseases, clinical symptoms of irreversible organ damage develop during pregnancy and could be prevented already before birth (Eddleman et al. 1996; Surbek et al. 1999). Human therapy with autologous CB stem cells harvested during pregnancy and genetically modified ex vivo has not yet been attempted. In more than 20 in utero transplantations of human fetuses which have been performed to date, bone marrow or fetal liver and thymus were used as sources of allogeneic stem cells. So far, success has been limited to recipients with immunodeficiency disorders (Touraine et al. 1989, 1992; Flake et al. 1996; Wengler et al. 1996). In other diseases such as hemoglobinopathies, engraftment has not been achieved either due to immunological intolerance or to occupation of bone marrow cavities with hematopoietic tissue (Slavin et al. 1992; Westgren et al. 1996). Since reduction of bone marrow cellularity by chemotherapy is not feasible in utero, improvement of repopulation with genetically modified cells may depend on methods to increase their number. Indeed, experiments in mice have demonstrated that engraftment of allogeneic cells without myelosuppressive conditioning can be achieved with a high dose of transplanted stem cells (Sykes et al. 1997).

The extensive in vitro proliferation of preterm CB progenitors in long-term cultures, as described in this work, may facilitate and broaden the clinical application of these cells. In the human fetus to be transplanted in utero with autologous CB progenitor cells, sufficient cell numbers may offer a feasible and promising outlook for treatment of a variety of hereditary disorders, not limited to immunodeficiencies and bone marrow failure syndromes. The biological properties of preterm CB, namely abundance of early and committed progenitors with high responsiveness to growth factors and ex vivo expansion and transduction potential equal to term CB, as well as high efficiency of purification, cryopreservation and recovery of viable $CD34^+$ cells described in our previous report (Surbek et al. 1998), are such that handling of fetal CB stem cells for transplantation should be as feasible as of term CB.

Acknowledgements. We thank Verena dalle Carbonare, Wendy Jansen and Sylvia Sendelov for technical assistance. This work has been supported by grants from the Swiss National Science Foundation 32-045926.95 and 4037-055157, Stiftung Basler Nabelschnurprojekt (Swisscord), Senglet Stiftung and Stiftung der Basler Chemischen Industrie (to A.L-W.).

References

Bordignon C, Notarangelo L, Nobili N, Ferrari G, Casorati G, Panina P, Mazzolari E, Maggioni D, Rossi C, Servida P ea (1995) Gene therapy in peripheral blood lymphocytes and bone marrow for ADA-immunodeficient patients. Science 270:470–475

Broxmeyer H, Douglas G, Hangoc G, Cooper S, Bard J, English D, Arny M, Thomas L, Boyse E (1989) Human umbilical cord blood as a potential source of transplantable hematopoietic stem/progenitor cells. Proc Natl Acad Sci USA 86:3828–3832

Broxmeyer H, Gluckman E, Auerbach A, Douglas G, Friedman H, Cooper S, Hangoc G, Kurtzberg J, Bard J, Boyse E (1990) Human umbilical cord blood: a clinically useful source of transplantable hematopoietic stem/progenitor cells. Int J Cell Cloning 8:76–89

Cairo M, Wagner J (1997) Placental and/or umbilical cord blood: an alternative source of hematopoietic stem cells for transplantation. Blood 90:4665–4678

Cardoso A, Li M-L, Batard P, Hatzfeld A, Brown E, Levesque J-P, Sookdeo H, Panterne B, Sansilvestri P, Clark S, Hatzfeld J (1993) Release from quiescence of CD34+CD38- human umbilical cord blood reveals their potentiality to engraft adults. Proc Natl Acad Sci USA 90:8707–8711

Civin C, Banquerigo M, Strauss L, Loken M (1987) Antigenic analysis of hematopoiesis. VI. Flow cytometric characterization of My-10-positive progenitor cells in normal human bone marrow. Exp Hematol 15:10–17

Clapp D, Baley J, Gerson S (1989) Gestational age-dependent changes in circulating hematopoietic stem cells in newborn infants. J Lab Clin Med 113:422–427

Conneally E, Eaves C, Humphries R (1998) Efficient retroviral-mediated gene transfer to human cord blood stem cells with in vivo repopulating potential. Blood 91:3487–3493

Costello E, Buetti E, Munoz M, Diggelmann H, Thali M (2000) Gene transfer into stimulated and un-stimulated T lymphocytes by HIV-1 derived lentiviral vectors. Gene Ther 7:596–604

Eddleman K, Chervenak F, George-Siegel P, Migliaccio G, Migliaccio A (1996) Circulating hematopoietic stem cell populations in human fetuses:

implications for fetal gene therapy and alterations with in utero red cell transfusion. Fetal Diagn Ther 11:231–240

Ehlers S, Smith K (1991) Differentiation of T cell lymphokine gene expression: the in vitro acquisition of T cell memory. J Exp Med 173:25–36

Emerson S, Sieff C, Wang E, Wong G, Clark S, Nathan D (1985) Purification of fetal hematopoietic progenitors and demonstration of recombinant multipotential colony-stimulating activity. J Clin Invest 76:1286–1290

Flake A, Zanjani E (1997) In utero hematopoietic stem cell transplantation. JAMA 278:932

Flake A, Harrison M, Adzick N, Zanjani E (1986) Transplantation of fetal hematopoietic stem cells in utero: the creation of hematopoietic chimeras. Science 233:776–778

Flake A, Roncarolo M, Puck J, Almeida-Porada G, Evans M, Johnson M, Abella E, Harrison D, Zanjani E (1996) Treatment of X-linked severe combined immunodeficiency by in utero transplantation of paternal bone marrow. N Engl J Med 335:1806–1810

Fleischman R, Mintz B (1979) Prevention of genetic anemias in mice by microinjection of normal hematopoietic stem cells into the fetal placenta. Proc Natl Acad Sci USA 76:5736–5740

Gluckman E, Broxmeyer H, Auerbach A, Friedman H, Douglas G, Devergie A, Esperou H, Thierry D, Socie G, Lehn P et al (1989) Hematopoietic reconstitution in a patient with Fanconi's anemia by means of umbilical-cord blood from an HLA-identical sibling. N Engl J Med 321:1174–1178

Gluckman E, Rocha V, Boyer-Chammard A, Locatelli F, Arcese W, Pasquini R, Ortega J, Souillet G, Ferreira E, Laporte J, Fernandez M, Chastang C (1997) Outcome of cord-blood transplantation from related and unrelated donors. Eurocord Transplant Group and the European Blood and Marrow Transplantation Group. N Engl J Med 337:373–381

Hao Q-L, Shah A, Thiemann F, Smogorzewska E, Crooks G (1995) A functional comparison of CD34+CD38-cells in cord blood and bone marrow. Blood 86:3745–3753

Harris D, Schumacher N, Locascio J, Besencon F, Olson G, DeLuca D, Shenker L, Bard J, Boyse E (1992) Phenotypic and functional immaturity of human umbilical cord blood T lymphocytes. Proc Natl Acad Sci USA 89:10006–10010

Harrison M, Slotnick R, Crombleholme T, Golbus M, Tarantal A, Zanjani E (1989) In utero transplantation of fetal liver haemopoietic stem cells in monkeys. Lancet 2:1425–1427

Hogge D, Lansdorp P, Reid D, Gerhard B, Eaves C (1996) Enhanced detection, maintenance, and differentiation of primitive human hematopoietic cells in cultures containing murine fibroblasts engineered to produce human steel factor, interleukin-3, and granulocyte colony-stimulating factor. Blood 88:3765–3773

Hunt P, Li Y, Nichol J, Hokom M, Bogenberger J, Swift S, Skrine J, Hornkohl A, Lu H, Clogston C, Merewether L, Johnson M, Parker V, Knudten A, Farese A, Hsu R, Garcia A, Stead R, Bosselmann R, Bartley T (1995) Purification and biologic characterization of plasma-derived megakaryocyte growth and development factor. Blood 86:540–547

Jones H, Nathrath M, Thomas P, Edelman P, Rodeck C, Linch D (1994) The effects of gestation of circulating progenitor cells. Br J Haematol 87:637–639

Kohn D, Weinberg K, Nolta J, Heiss L, Lenarsky C, Crooks G, Hanley M, Annett G, Brooks J, El-Khoureiy A, Lawrence K, Wells S, Moen R, Bastian J, Williams-Herman D, Elder M, Wara D, Bowen T, Hershfield M, Mullen C, Blaese R, Parkman R (1995) Engraftment of gene-modified umbilical cord blood cells in neonates with adenosine deaminase deficiency. Nat Med 1:1017–1023

Lansdorp P, Dragowska W, Mayani H (1993) Ontogeny-related changes in proliferative potential of human hematopoietic cells. J Exp Med 178:787–791

Larochelle A, Vormoor J, Hanenberg H, Wang J, Bhatia M, Lapidot T, Moritz T, Murdoch B, Xiao X, Kato I, Williams D, Dick J (1996) Identification of primitive human hematopoietic cells capable of repopulating NOD/SCID mouse bone marrow: implications for gene therapy. Nat Med 2:1329–1337

Migliaccio G, Baiocchi M, Hamel N, Eddleman K, Migliaccio A (1996) Circulating progenitor cells in human ontogenesis: response to growth factors and replating potential. J Hematother 5:161–170

Miyoshi H, Smith KA, Mosier DE, Verma IM, Torbett BE (1999) Transduction of human CD34+ cells that mediate long-term engraftment of NOD/SCID mice by HIV vectors. Science 283:682–686

Moore M, Hoskins I (1994) Ex vivo expansion of cord-blood derived stem cells and progenitors. Blood Cells 20:468–481

Naldini L, Blomer U, Gallay P, Ory D, Mulligan R, Gage F, Verma I, Trono D (1996) In vivo gene delivery and stable transduction of nondividing cells by a lentiviral vector. Science 272:263–267

Péault B (1996) Hematopoietic stem cell emergence in embryonic life: developmental hematology revisited. J Hematother 5:369–378

Piacibello W, Sanavio F, Garetto L, Severino A, Bergandi D, Ferrario J, Fagioli F, Berger M, Aglietta M (1997) Extensive amplification and self-renewal of human primitive hematopoietic stem cells from cord blood. Blood 89:2644–2653

Rubinstein P, Carrier C, Scaradavou A, Kurtzberg J, Adamson J, Migliaccio A, Berkowitz R, Cabbad M, Dobrila N, Taylor P, Rosenfield R, Stevens C

(1998) Outcomes among 562 recipients of placental-blood transplants from unrelated donors. N Engl J Med 339:1565–1577

Sautois B, Fillet G, Beguin Y (1997) Comparative cytokine production by in vitro stimulated mononucleated cells from cord blood and adult blood. Exp Hematol 25:103–108

Shields L, Andrews R (1998) Gestational age changes in circulating CD 34+ hematopoietic stem/progenitor cells in fetal cord blood. Am J Obstet Gynecol 178:931–937

Slavin S, Naparstek E, Ziegler M, Lewin A (1992) Clinical application of intrauterine bone marrow transplantation for treatment of genetic diseases – feasibility studies. Bone Marrow Transplant 9:189–190

Surbek D, Holzgreve W, Jansen W, Heim D, Garritsen H, Nissen C, Wodnar-Filipowicz A (1998) Quantitative immunophenotypic characterization, cryopreservation, and enrichment of second and third trimester human fetal cord blood hematopoietic stem cells (progenitor cells). Am J Obstet Gynecol 179:1228–1233

Surbek D, Gratwohl A, Holzgreve W (1999) In utero hematopoietic stem cell transfer: current status and future strategies. Eur J Obstet Gynecol Reprod Biol 85:109–115

Sutherland H, Lansdorp P, Henkelman D, Eaves A, Eaves C (1990) Functional characterisation of individual human hematopoietic stem cells cultures at limiting dilution of supportive marrow stromal layers. Proc Natl Acad Sci USA 87:3584–3588

Sykes M, Szot G, Swenson K, Pearson D (1997) Induction of high levels of allogeneic hematopoietic reconstitution and donor-specific tolerance without myelosuppressive conditioning. Nat Med 3:783–787

Tavassoli M (1991) Embryonic and fetal hemopoiesis: an overview. Blood Cells 17:269–281

Thilaganathan B, Nicolaides K, Morgan G (1994) Subpopulations of CD34-positive haemopoietic progenitors in fetal blood. Br J Haematol 87:634–636

Touraine J, Raudrant D, Royo C, Rebaud A, Roncarolo M, Souillet G, Philippe N, Touraine F, Betuel H (1989) In utero transplantation of stem cells in bare lymphocyte syndrome. Lancet 1:1382

Touraine J, Raudrant D, Rebaud A, Roncarolo M, Laplace S, Gebuhrer L, Betuel H, Frappaz D, Freycon F, Zabot M et al (1992) In utero transplantation of stem cells in humans: immunological aspects and clinical follow-up of patients. Bone Marrow Transplant Suppl 1:121–126

Uchida N, Sutton R, Friera A, He D, Reitsma M, Chang W, Veres G, Scollay R, Weissman I (1998) HIV, but not murine leukemia virus, vectors mediate high efficiency gene transfer into freshly isolated G0/G1 human hematopoietic stem cells. Proc Natl Acad Sci USA 95:11939–11944

Udomsakdi C, Lansdorp P, Hogge D, Reid D, Eaves A, Eaves C (1992) Characterization of primitive hematopoietic cells in normal human peripheral blood. Blood 80:2513–2521

van Hennik P, Verstegen M, Bierhuizen M, Limon A, Wognum A, Cancelas J, Barquinero J, Ploemacher R, Wagemaker G (1998) Highly efficient transduction of the green fluorescent protein gene in human umbilical cord blood stem cells capable of cobblestone formation in long-term cultures and multilineage engraftment of immunodeficient mice. Blood 92:4013–4022

Wang J, Doedens M, Dick J (1997) Primitive human hematopoietic cells are enriched in cord blood compared with adult bone marrow or mobilized peripheral blood as measured by the quantitative in vivo SCID-repopulating cell assay. Blood 89:3919–3924

Wengler G, Lanfranchi A, Frusca T, Verardi R, Neva A, Brugnoni D, Giliani S, Fiorini M, Mella P, Guandalini F, Mazzolari E, Pecorelli S, Notarangelo L, Porta F, Ugazio A (1996) In-utero transplantation of parental CD34 haematopoietic progenitor cells in a patient with X-linked severe combined immunodeficiency (SCIDXI). Lancet 348:1484–1487

Westgren M, Ringden O, Eik-Nes S, Ek S, Anvret M, Brubakk A, Bui T, Giambona A, Kiserud T, Kjaeldgaard A, Maggio A, Markling L, Seiger A, Orlandi F (1996) Lack of evidence of permanent engraftment after in utero fetal stem cell transplantation in congenital hemoglobinopathies. Transplantation 61:1176–1179

Wodnar-Filipowicz A, Tichelli A, Zsebo K, Speck B, Nissen C (1992) Stem cell factor stimulates the in vitro growth of bone marrow cells from aplastic anemia patients. Blood 79:3196–3202

Wyrsch A, dalle Carbonare V, Jansen W, Chklovskaia E, Nissen C, Surbek D, Holzgreve W, Tichelli A, Wodnar-Filipowicz A (1999) Umbilical cord blood from preterm human fetuses is rich in committed and primitive hematopoietic progenitors with high proliferative and self-renewal capacity. Exp Hematol 27:1338–1345

Zanjani E, Pallavicini M, Ascensao J, Flake A, Langlois R, Reitsma M, MacKintosh F, Stutes D, Harrison M, Tavassoli M (1992) Engraftment and long-term expression of human fetal hemopoietic stem cells in sheep following transplantation in utero. J Clin Invest 89:1178–1188

10 Tolerance Induction Post In Utero Stem Cell Transplantation

M.J. Cowan, S.-H. Chou, A.F. Tarantal

10.1 Introduction

Allogeneic bone marrow (BM) stem cell transplantation in childhood is used to cure a variety of inherited diseases including severe immunodeficiency diseases, hemoglobinopathies, and some storage diseases (O'Reilly et al. 1984; Cowan 1991). For all of these disorders, except most types of severe combined immunodeficiency disease (SCID), immunosuppressive chemotherapy is required to prevent graft rejection. Also, in order to obtain sufficient erythroid and/or myeloid engraftment, marrow ablative chemotherapy is necessary to "make space" for the donor hematopoietic stem cells (HSC). This conditioning is associated with significant morbidity and a mortality of 10–15%. Finally, for some diseases (e.g., Hurler's mucopolysaccharidosis), significant organ damage has already occurred by the time the diagnosis is made postnatally (Hobbs 1988). The need for immunosuppression can be eliminated if the transplant is done sufficiently early in utero when the fetus is unable to reject allogeneic donor HSC.

10.2 In Utero Transplantation in Animals and Humans

Results of studies of in utero HSC transplantation in mice, sheep, cats, and non-human primates make it clear that transplantation during the first trimester or early second trimester is essential to avoid rejection (Harrison et al. 1989; Roodman et al. 1991; Cowan et al. 1996). Also, both fetal liver and adult BM can successfully engraft (Cromblehome et al. 1990; Srour et al. 1992; Carrier et al. 1995, 1996; Cowan et al. 1996). Unfortunately, a consistent finding in those models (mice, sheep, and monkeys) in which the recipient's hematopoietic system is intact (i.e., in which there is no stem cell or progenitor cell defect) has been a low percentage (often <1 %) of engrafted donor cells, inadequate to cure most candidate diseases (Srour et al. 1992; Flake and Zanjani 1993; Carrier et al. 1995, 1996; Cowan et al. 1996).

There have been more than three dozen in utero HSC transplants in humans (Cowan and Golbus 1994; Jones et al. 1996; Flake and Zanjani 1999). Fetal liver was first used successfully for a fetus with the bare lymphocyte syndrome (Touraine 1991). Subsequent in utero transplants have used fetal liver for α- and β-thalassemia, sickle cell anemia, Hurler's disease and Nieman-Pick disease (Touraine 1991; Westgren et al. 1996). While there has been evidence of engraftment in some but not all (Westgren et al. 1996), because of a very low percentage of donor cells, there has been no clinical benefit. Parental adult marrow has engrafted in two fetuses with α-thalassemia (Cowan and Golbus 1994; Hayward et al. 1998) and several fetuses with SCID (Flake et al. 1996; Wengler et al. 1996; Gil et al. 1999), but only the children with SCID have benefited clinically. The low percentage of engrafted donor cells is the greatest limitation to using in utero stem cell transplantation to treat the majority of inherited diseases in which there is no inherent survival advantage for donor HSC or committed progenitors.

10.3 Issues Associated with In Utero Transplantation

Based on the results of studies in animals and humans, it is apparent that there are several issues that need to be addressed before successful in utero transplantation can be accomplished. They include: the ability of the fetus to reject donor cells, the susceptibility of the fetus to tolerance induction, space, the fetal environment, and the source of donor cells.

10.3.1 Immunocompetence of the Fetal Recipient

Based on studies of the ontogeny of the immune system in humans and animals, the results of fetal tissue (thymus and liver) transplants in children with SCID, and the experience with fetal transfusion therapy, it appears that a critical gestational age exists beyond which T-cell immunity is sufficient to reject donor HSC (Cowan and Golbus 1994; Flake and Zanjani 1999). The precise time is not clear, although it appears to be early in the second trimester at ~14 weeks' gestational age for human fetuses. In rhesus monkey, sheep, and mouse models of in utero transplantation there also appears to be a fetal age at which allogeneic donor cells will engraft (Flake and Zanjani 1993; Cowan et al. 1996). However, in spite of numerous efforts to engraft fetal recipients at these ages, only a relatively small number of donor cells durably persist and in only limited circumstances does tolerance develop long term towards the donor cells.

10.3.2 Transplant Rejection and Tolerance Induction

Tolerance to self antigens is a critical developmental phenomenon that is essential to the survival of complex organisms. A great deal of interest has focused on the development of self-tolerance in order to understand the pathogenesis of autoimmune diseases (Kruisbeek and Amsen 1996; de St Groth 1998). This information has been useful in pursuing the mechanisms of transplant rejection and tolerance induction towards allogeneic cells and tissues (Sayegh and Turka 1998; Rossini et al. 1999). The proposed models for T-cell tolerance induction have had to account for both the immature developing thymocyte as well as the mature T cell that has escaped intrathymic negative selection. It is currently believed that self-tolerance begins with T cells that are maturing in the thymus and express an antigen receptor with high affinity for a self-antigen. These T cells are negatively selected to die within the thymus when the T cell receptor (TCR) is engaged (Kruisbeek and

AMsen 1996; Fazekas d St Groth 1998). It is believed that marrow-derived antigen-presenting cells (APC) are important in this phase. However, this process does not eliminate all self-reactive thymocytes, and not all self peptides are likely to be presented to T cells during their maturation. Thus, peripheral mechanisms must exist which result in either T cell apoptosis or non-responsiveness (anergy), and serve the purpose of not only maintaining tolerance, but modulating the degree and duration of an immune response (Sayegh and Turka 1998; Rosini et al. 1999).

Peripheral deletion is regulated in part by interaction between Fas (CD95) and its ligand (FasL) (Crispe 1994). In the adult mouse model it has been shown that FasL expression on injected donor marrow cells was essential for tolerance induction and that microchimerism alone was not sufficient (George et al. 1998). However, there are other Fas-independent mechanisms for peripheral deletion (Zheng et al. 1995). Anergy occurs principally in the extrathymic periphery, although anergy clearly can take place in the thymus in newly mature T cells (Ramsdell and Fowlkes 1990). There are also situations in which neither deletion nor anergy appear to be present and evidence exists for suppression or veto cells of lymphoid origin (Qin et al. 1993; Thomas et al. 1994; Martin 1996).

The underlying mechanism for allogeneic graft rejection involves T cell activation. There is ample evidence that a 2-signal process is required to activate T cells. For cytotoxic T cells (CD8), the first signal is mediated via the TCR in contact with antigen that is presented to the TCR in the context of the major histocompatibility complex (MHC) (Sayegh and Turka 1998). The second signal comes from a professional [e.g., dendritic cell (DC) or macrophage], which in turn has been activated via interaction with the CD40 ligand of the T helper cell (Ridge et al. 1998). CD40 ligation results in DC secretion of cytokines such as interleukin (IL)-1β, tumor necrosis factor alpha (TNF-α) and IL-12. It also increases DC expression of MHC class II, adhesion molecules and the co-stimulatory molecules, B7.1 (CD80) and B7.2 (CD86) (Caux et al. 1994; Karmann et al. 1995). Interfering with CD40 ligand/CD40 interaction significantly hinders T cytotoxic cell maturation by blocking T cell-DC interaction (Bennett et al. 1998; Schoenberger et al. 1998). T cells generate an immune response when they interact with antigen/MHC (signal 1) along with the second signal, but they die without the second signal. The effectiveness of blocking the second signal, i.e., CD40 ligand/CD40 interaction, in inducing tolerance to solid organs has been shown in mice (Larsen et al. 1996; Honey et al. 1999) and more recently, in a monkey model of kidney transplantation (Kirk et al. 1999). When signal 1 is blocked, e.g., with cyclosporin (CSA), the immune response is limited, but as soon as CSA administration is stopped, T cells become activated and the transplanted tissue is rejected. In the monkey study (Kirk et al. 1999) the second signal (CD40L/CD40) was blocked with an anti-CD40L monoclonal antibody resulting in permanent tolerance induction in the majority of animals. However, when CSA was added to the regimen (blocking the TCR-mediated signaling pathway) permanent tolerance no longer developed, indicating the importance of the CD40L/CD40 pathway not only in activation, but also in tolerance induction.

Another critical T cell activation pathway is via CD28, which is expressed on resting T cells, and its ligands on activated professional APC, B7.1 and B7.2 (Harding et al. 1992; Freeman et al. 1993). Following signal 1 via engagement of the TCR, CD28 ligation results in IL-2 secretion, IL-2 receptor upregulation, and proliferation. B7.1 and B7.2 also bind CTLA-4 (CD152), an immunoglobulin superfamily member that has homology to CD28. CTLA-4 is upregulated within hours of T cell activation and delivers a negative signal to activated T cells (Green et al. 1994; Walunas et al. 1994). CTLA4-Ig is a synthetic molecule combining the extracellular domain of CTLA-4 and the constant region of IgG1. CTLA4-Ig binds to B7.1 and B7.2 with a higher affinity than CD28. Blockade of B7.1/B7.2 interaction with CD28 by monoclonal antibodies or CTLA4-Ig significantly prolongs allograft survival and prevents graft versus host disease (GVHD) in a variety of animal models (Lenschow et al. 1992; Lin et al. 1993; Blazar et al. 1994).

Studies in solid organ and BM transplant models in mice are particularly interesting in terms of designing experiments to understand tolerance induction following in utero stem cell transplantation. In the solid organ transplant experience, in particular with liver transplants, donor multilineage microchimerism has been reported, presumably originating from the donor liver (Thomson et al. 1995). Studies of liver (Lu et al. 1995) and cardiac (Thomson et al. 1995) transplantation in mice have revealed that microchimerism develops in the former along with tolerance, while no donor hematopoietic cells can be detected in the latter situation in which tolerance is not typically induced. Marrow-derived DCs in the spleen, blood, and thymus are thought to be critically important in tolerance induction in these models. It has been proposed that because the BM-derived DC progenitor expresses MHC class II molecules but lacks the expression of critical T cell co-stimulatory molecules, B7.1 and B7.2, host T cell interaction with these donor cells from the transplanted liver results in anergy (Thomson et al. 1995). In the murine heart transplant model, CTLA4-Ig results in prolonged but not permanent graft survival (Lin et al. 1993; Lenschow et al. 1992). However, if donor splenocytes (even if irradiated) are administered at the time of transplant in the presence of systemic CTLA4-Ig, a permanent state of tolerance will develop in most animals (Sayegh et al. 1997). Donor DC precursors have been found in both human and animal models of liver transplantation and have been proposed as candidates for the induction of tolerance post-transplant. It is also possible that host professional APC could present donor antigens such that tolerance could also be induced, as has been shown for the Mls system (Ramsdell and Fowlkes 1990), although this seems more likely to result in activation than tolerance induction.

10.3.3 Space

In hematopoietically intact fetal mouse models, using syngeneic or congenic donor cells from either fetal or adult sources, it is possible to durably engraft only a limited number of cells, generally less than 1% of the recipient cell population (Carrier et al. 1995, 1996). In the fetal mouse model only a single injection is technically possible in contrast to adult mice (Stewart et al. 1993) and larger animal models (Flake and Zanjani 1993; Tarantal et al. 2000) and humans (Hayward et al. 1998; Flake et al. 1996) in which multiple injections can be given. In adult mice, repeated injections of large numbers or a single extremely large number of syngeneic marrow cells results in a significant percentage of durably engrafted donor cells (Stewart et al. 1993; Ramshaw et al. 1995; Rao et al. 1997). It appears that competition for space in the adult recipient is less of an issue than in the fetus. This may vary, in particular, on the gestational age and site of hematopoiesis, i.e., liver vs BM. The circumstances under which significant donor cell engraftment occurs in utero have been in either fetal mice with an underlying hematopoietic stem cell defect (Fleischman and Mintz 1979; Archer et al. 1997; Blazar et al. 1998) or in human fetuses with SCID. However, in the latter situation, only small numbers of donor HSC engraft with commitment almost exclusively towards lymphoid progenitors (Stiehm et al. 1996). In contrast to what has been seen in these defective mouse models, the human SCID recipients only engraft with donor lymphocytes (Flake et al. 1996; Wengler et al. 1996; Gil et al. 1999). This is comparable to what is seen in postnatal HSC transplants for children with SCID who are not conditioned with myeloablative chemotherapy, i.e., only T cells and occasionally B cells are of donor origin, while the myeloid and erythroid progenitor cell lines remain of host origin (Merino et al. 1993; Brady et al. 1996;). Several in utero HSC transplants in humans have been done for hemoglobinopathies, specifically α- and β-thalassemia and sickle cell disease in which a survival advantage for engrafted donor erythroid cells would be expected. While there has been occasional evidence of engraftment, no surviving child has had a sufficient number of donor cells to alter the course of the disease (Cowan and Golbus 1994; Touraine 1996; Westgren et al. 1996; Hayward et al. 1998).

10.3.4 The Fetal Environment

In the human, the initial source of HSC is the yolk sac (Kelemen et al. 1979; Christensen 1989). Migration of HSC into the hepatic primordia for establishment of hematopoiesis occurs during the 5th–6th week. The liver becomes the major source of blood cells, with a peak in production in the 3rd month and subsequent decline after the 7th month. BM becomes an active site of hematopoiesis at 10–11 weeks' gestation. BM stromal matrix develops approximately 2 weeks prior to accepting the immigration of blood-born HSC, which colonize primitive marrow stroma and divide and differentiate to form extrasinusoidal foci of hematopoiesis. Maturation of hematopoiesis in the fetus has been studied with respect to expression of maturation markers and growth of HSC isolated from fetal tissue (Jordan et al. 1990; Ogawa et al. 1993; Muench et al. 1994; Tocci et al. 1995). Studies in the monkey have shown comparable hematopoietic events at comparable developmental time periods (Sharma et al. 1991; Tarantal 1993; Tarantal and Cowan 1999).

HSC from human fetal liver and cord blood differ from adult marrow progenitors in cycling rates in that they are a constant and rapidly expanding pool, and have accelerated in vitro maturation times (Christensen 1989; Hahn et al. 1994). They can also react to cytokines differently than adult marrow HSC. For example, they are highly resistant to the growth inhibitory effects of interferon (Hahn et al. 1994). Stromal cells from fetal liver and cord blood have also been compared with adult marrow stroma in terms of the ability to support in vitro HSC growth, with differences being found as a function of age (Slaper-Cortenbach et al. 1987; Van den Heuvel et al. 1991).

There is evidence in sheep that fetal liver HSC migrate to the BM rather than marrow hematopoiesis developing de novo. Thus, the timing of the transplant is critical, not only from the standpoint of fetal immunocompetence, but in terms of the site of active hematopoiesis at that time. The optimal site of engraftment is unknown (liver vs BM), but whether or not donor cells that engraft in the fetal liver are capable of migrating to the BM could determine ultimate efficacy. In addition, competition for engraftment with host HSC may be quite different in the fetal liver vs the fetal marrow. Finally, investigators have suggested that histocompatibility between HSC and stroma may be essential for optimal engraftment (Flake and Zanjani 1999). It is difficult to know the seriousness of this problem, since engraftment across MHC barriers can occur in immunodeficient and stem cell-defective animal models (SCID, NOD-SCID, W41, and Wv mice) and haplocompatible postnatal transplants in humans (Dror et al. 1993; Aversa et al. 1994; de la Selle et al. 1996).

10.3.5 The Source of Donor Cells

The optimal source of HSC for transplantation is still unknown. With respect to "quality" or engraftment capability, there is evidence that immature sources of HSC, that is, cord blood or fetal liver, are better than adult BM HSC (Fleischman and Mintz 1984; Papayannopoulou et al. 1993; de la Selle et al. 1996; Tarantal et al. 2000). With respect to cell numbers and availability, adult BM or recruited peripheral blood stem cells (PBSC) are attractive sources, since the number of cells in fetal and even cord blood is limited. This may be especially important for in utero transplants in which booster transplants may be necessary to enhance engraftment postnatally. PBSC are particularly interesting because of evidence that these cells not only have long-term repopulating capacity, but that they have significantly higher cloning efficiencies than adult marrow (Bodine et al. 1996; Yamamoto et al. 1996). The ability of allogeneic (matched or haplocompatible) PBSC to durably engraft in humans has been documented (Bensinger et al. 1995).

10.4 Studies at the University of California San Francisco

10.4.1 In Utero Haplocompatible Parental BMT in Rhesus Monkeys

10.4.1.1 T Cell-Depleted Marrow

We established a model of in utero stem cell transplantation in the fetal rhesus monkey (Cowan et al. 1996). Rhesus monkeys (Macaca mulatta) were housed at the California Regional Primate Research Center (CRPRC) and used under protocols approved by the University of California Animal Use and Care Committee. Established aseptic ultrasound-guided procedures were used for all in utero collections (chorionic villous sampling, fetal blood collection, and fetal liver aspirates) and intraperitoneal fetal injections (Tarantal and Hendrickx 1988; Tarantal 1990).

We used a polymerase chain reaction (PCR) assay for the rhesus Y chromosome, sensitive to 0.01%, to determine sex from GD 30 chorionic villous samples, and engraftment in female recipients of male (sire) marrow. The PCR reaction was a modification of that previously reported by Reitsma et al. (1993). This reaction produces a 174-bp band which is specific for the male Rhesus monkey and Olive baboon. These primers do not amplify human male DNA, nor do they amplify female Rhesus monkey or Olive baboon.

We wanted to test the hypothesis that earlier in gestation the rate and degree of engraftment would be better because of less host resistance and fewer competing host HSC. We compared the engraftment rate of T cell-depleted (TCD) haplocompatible parental marrow in two groups of recipients based upon their mean gestational ages, 44±2 days versus 61±4 days (normal gestation is 165 days). Multiple marrow aspirates

(1–2 cc/aspiration) from the posterior iliac crests in anesthetized sires (for female fetal recipients) were collected in RPMI 1640 with 100 U/ml heparin. The marrow was T cell depleted using soybean agglutinin and 2 cycles of sheep erythrocyte rosetting, as previously described (Cowan et al. 1985). The TCD marrow was concentrated to 200 λ and injected intraperitoneally (IP) in utero under ultrasound guidance. The T cell depletion and transplant occurred within 24 h of the marrow harvest. Engraftment rates in both groups of animals were similar (66% and 80%, respectively). Donor cells were found in BM, blood, spleen and liver. There was no evidence of GVHD in any of the engrafted animals (up to 3 years' follow-up). However, the percentage of donor cells remained low (<0.1%) regardless of the gestational age of the recipient, suggesting either inadequate homing and/or space.

10.4.1.2 Tolerance Induction

We evaluated the tolerance of engrafted animals using several approaches (Cowan et al. 1996; Mychaliska et al. 1997). In the majority of experiments there was no proliferative response of recipient cells to donor targets, with normal responses to third-party cells similar to what we had found in children with SCID post-haplocompatible parental BMT (Merino et al. 1993). We also found absent cell-mediated lymphocytotoxicity (CML) responses of recipient cells to donor targets in two separate experiments (Cowan et al. 1996). We performed kidney transplants in four animals in which no immunosuppression was given (Mychaliska et al. 1997). In two controls in which the donor was either unrelated or was a parent (haplocompatible), the kidneys were acutely rejected within 7 days. The third animal was engrafted in utero with paternal marrow. There was no clinical evidence of rejection until 9 weeks. The fourth recipient was also engrafted with paternal cells. At 8 months post-kidney transplant on no immunosuppression and with no clinical evidence of rejection, the animal was electively killed. While the kidney did show histologic evidence of acute and chronic rejection, it was less severe than that seen in the controls at 1 week post-kidney transplant. These results indicate that in this non-human primate model, TCD parental BM cells can durably engraft in first or early second trimester fetal recipients and, at least in some animals, induce a partial state of tolerance.

Table 1. Engraftment following multiple in utero boosts

Monkey	GD	Number of boosts[a]	Liver[c]	Blood[c]	Marrow[c]
28549	48	1@GD90	Yes	?[b]/0	Yes
28557	46	3@GD95,110,115	Yes	?/?	Yes
28561	44	5@GD95,110,120, 125,130	Yes	Yes/Yes	Yes

[a] $3-10\times10^6$ T cell depleted donor marrow per injection.
[b] Borderline positive.
[c] Engraftment in liver pre-boosts; in blood pre-/post-boosts; in marrow post-natally.

10.4.1.3 Multiple In Utero Injections

We also did an experiment to determine whether multiple in utero boosts were feasible and effective. Three female fetuses were transplanted on gestational days (GD) 44–48 with paternal TCD marrow (Table 1). One animal received five boosts with TCD marrow between GD 95 and 130 at 5- to 10-day intervals; one animal was boosted on GD 95, 110 and 115, and one animal received a second transplant on GD 90. Liver and blood samples were obtained in utero prior to the initiation of boosts for engraftment. All animals were engrafted in the fetal liver, and post-natally in the marrow. Marrow samples were analyzed at 1 year of age by karyotyping. We found 0–2% donor cells, suggesting that there was no significant benefit of multiple in utero boosts in this animal model, even when the estimated dose of donor cells per injection was $\sim2\times10^8$/kg recipient body weight.

10.4.1.4 Cytokine-Recruited PBSC Transplants

We have performed two in utero HSC transplants using cytokine-mobi-lized PBSC. The recipients were female fetuses at GD 55 58. The sires were treated with daily subcutaneous injections of stem cell factor (SCF) (25 µg/kg) and G-CSF (100 µg/kg), and the apheresis was done on day 10. The cells were positively selected for CD34 expression with Dynal CD34-conjugated magnetic beads (Wagnum et al. 1994), and TCD with anti-CD2 monoclonal antibody conjugated to Dynal beads, leaving a final sample of 4.3×10^6 cells, which was 66% CD34$^+$ and <0.1% CD2$^+$. These were injected IP into the fetal recipient. The second experimental apheresis (on day 9) yielded 6×10^{10} cells (Fig. 1). One-third of the cells were CD34 positively selected and T cell depleted

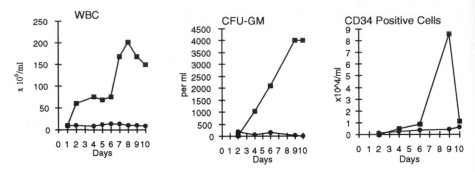

Fig. 1. Mobilization of CD34$^+$ cells with SCF/G-CSF (■) in adult rhesus monkey. Nine daily injections were given, following which the animal was apheresed on day 10. ● represents control animal

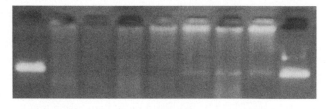

Fig. 2. Engraftment post in utero transplantation of recruited PBSC. Blood (*B*) and marrow (*M*) from monkeys 138 and 154, 1 and 3 months of age, respectively. Monkey 138 received 4.3×10^6 and monkey 154 received 28×10^6 CD34-enriched cells on gestational days 55 and 58, respectively. *C*, male controls; *F*, female controls

Table 2. CD34$^+$ enrichment of recruited PBSC from rhesus monkey

Cell number	CD2^a	CD34	CD34$^+$38$^-$	CFU-GM^b	BFU-E^b	CFU-MIX^b
Pre	2.4×10^{10}	54%	1.6%	ND	213	3	0
Post	28×10^6	0.1%	69%	16%	7075	900	725

ND, not done
^{a} Percentage of all cells.
^{b} Per 10^5 cells plated.

(Table 2). The resulting preparation consisted of 28×10^6 cells which were <0.1% CD2 and 70% CD34$^+$. Further characterization of these cells showed that a significant number were CD34$^+$38$^-$. CFU-GM and BFU-E were enriched 33- and 300-fold, respectively (Table 2). Engraftment at 3 months of age showed donor cells present in the blood and marrow of the second recipient (28×10^6 cells) and marrow of the first recipient, who received a lower cell dose (Fig. 2).

10.4.1.5 Xenogeneic Model of In Utero HSC Transplantation

We evaluated engraftment efficiency in a primate xenogeneic model of in utero HSC transplantation (Tarantal et al. 2000). Human PBSC were obtained by leukapheresis from a volunteer human male donor following 4 days of recombinant human granulocyte-colony stimulating factor (rhG-CSF) (5 μg/kg per day). The PBSC were enriched for CD34$^+$ cells using the Isolex300i cell separator. A fraction of the CD34$^+$ cells were then further T cell depleted resulting in two cell populations, one that was stem cell enriched (0.83% CD3$^+$ cells, 95% CD34$^+$; group 1) and one that was stem cell enriched and T cell depleted (<0.03%CD3$^+$ cells, 98% CD34$^+$; group 2).

Four fetal monkeys (*n*=2/group) received either 2 or 4 intraperitoneal (IP) injections (~5×10^6 cells per injection) via ultrasound guidance every other day over a 7-day period GD 50, 52, 54, and 56). One fetus in each group also received rhSCF (25 μg/kg) and rhG-CSF (10 μg/kg) IP post-transplant every 10 days from GD 60–150. Four healthy newborns were delivered at term, and blood and marrow were analyzed by PCR for the human Y chromosome at birth and monthly to 6 months of age. For some samples, progenitor assays were set up and individual erythroid and myeloid colonies analyzed for donor cell engraftment.

PCR results were positive for all four newborns in all specimens assessed, and FACS analysis for human CD45 in the marrow showed engraftment ranging from 0.1–1.7%. There was no evidence of GVHD in any of the animals.

These results indicate that multilineage engraftment of human PBSC can be achieved in the fetal rhesus recipient. The rhesus fetus appears to tolerate relatively high numbers of human CD3$^+$ cells without developing GVHD, although in this limited number of animals we were unable to show that increasing the number of donor T cells in the graft affected the degree of donor cell engraftment. Furthermore, we could find no effect of injecting cytokines into these recipients in terms of enhancing the growth advantage of the human cells over the rhesus HSC.

10.4.2 In Utero Mouse Model

10.4.2.1 Early Gestation Mice

We established a murine model of in utero transplantation in order to evaluate the role of MHC in engraftment efficacy and compare fetal liver to adult BM as sources of HSC in 11- to 13-day fetal recipients (Carrier et al. 1995, 1996; Hajdu et al. 1996). The congeneic model consisted of C57BL/6 (GPI-1a, Thy 1.1) recipients and congeneic C57BL/6 (GPI-1b, Thy 1.2) donors. The MHC-mismatched model used C57BL/6 (H-2^b, GPI-1b) donors and BALB/c (H-2^d, GPI-1a) recipients. To detect engraftment, we used a quantitative allele-specific (QAS) PCR assay for the murine Y chromosome which can detect male donor cells to 0.00001%. While we found some variability in the engraftment in various tissues in the congeneic and allogeneic mismatched animals, the overall durable engraftment rates were 50% and 45%, respectively. The percentage of donor cells varied from as low as 0.0001% in tissues (spleen or liver) to 0.6% in peripheral blood. In general, the percentage of donor cells was higher in blood and marrow than in liver or spleen. Interestingly, donor cells were found in BM in 80% of recipients of allogeneic mismatched fetal liver. In some of these animals, there was no other evidence of engraftment (Carrier et al. 1995).

10.4.2.2 Bone Marrow as a Source of HSC

Because there are significant practical limitations associated with fetal liver as a source of stem cells, we wanted to characterize the engraftment and tolerance-inducing capabilities of MHC-mismatched adult marrow in non-defective fetal recipients using the same congeneic and allogeneic pairs (Carrier et al. 1996). The overall engraftment rate (in the tissues and blood) was 44% in allogeneic and 44% in congeneic recipients. The QAS-PCR assay was used to estimate the degree of engraftment in the blood, liver, spleen and BM. The range of donor cells in the peripheral blood was 0.002–2.4%, with a median of 0.01%. Similarly, a low degree of engraftment was observed in the tissues with a range of 0.0003–0.4% (median 0.01%). There was no significant difference in the percentage of donor cells between allogeneic and congeneic recipients ($P>0.3$). Compared with our previous study using fetal liver, we could find no obvious differences with respect to engraftment rate or percentage of donor cells present in engrafted animals with the use of adult marrow HSC in this in utero transplant model. These results, in fact, were quite similar to what we found in the rhesus monkey model using parental TCD BM (Cowan et al. 1996).

10.4.2.3 Phenotype of Engrafted Donor Cells

Using FACS analysis we phenotyped peripheral blood from a fetal liver recipient with a relatively high degree of engraftment (Carrier et al. 1996). Donor T lymphocytes were detected in the peripheral blood. The QAS-PCR showed 3.5% donor cells and by FACS analysis, 6.5% were cells of donor origin, 80% of which were CD3-positive. To characterize the donor cells in the tissues, we performed QAS-PCR on c-kit and CD3 positively selected populations of splenocytes from two durably engrafted BM recipients at 26 months of age. In one mouse there was no evidence of donor cells in either the CD3 or c-kit$^+$ fractions. In the other mouse (which never had detectable circulating donor cells in the blood), the PCR signal in the c-kit$^+$ splenocytes (>95%) was strongly positive, while it was negative in the CD3$^+$ fraction. Quantitative analysis revealed ~0.05% donor c-kit$^+$ cells present in the preparation.

10.4.2.4 Tolerance Induction

To date, we have relied on donor skin graft acceptance as the measure of tolerance in murine recipients. In our initial study of 22 female recipients of allogeneic MHC-mismatched fetal liver HSC, 3 animals with permanent donor skin grafts all had circulating donor cells, although 1 of the animals became negative in the blood shortly after the skin graft was placed. Interestingly, donor cells from one of these animals was phenotyped (see above), and donor T cells were present. None of the BM recipients permanently accepted skin grafts including an animal with 2% circulating donor cells (none of which were $CD3^+$). Finally, in order to establish whether recruitment of donor cells into the peripheral circulation in non-tolerant recipients could induce or alter tolerance, we selected 5 mice, previously transplanted in utero with fetal liver, which had no prior evidence of engraftment on multiple studies in the blood, liver or spleen and had previously rejected donor skin grafts. They were injected with SCF/G-CSF for 7 days and on day 8 had donor skin grafts placed. Three showed the presence of donor cells in the peripheral blood following injection with cytokines, but all 5 recipients rejected donor and third-party skin grafts within 14 days.

Our results indicate that relatively low numbers of engrafted donor cells are necessary to induce a state of at least partial donor skin graft tolerance Also, circulating donor cells may not be necessary for maintenance of and are not sufficient for the induction of skin graft tolerance. Finally, some engrafted animals will retain the ability to reject donor skin grafts in spite of relatively large numbers (2%) of circulating donor cells. Our phenotyping data indicate that at least some animals can develop circulating donor T cells. We believe this may be critical in the development of tolerance in this model.

10.4.2.5 Postnatal Boosts

We injected donor allogeneic MHC-mismatched fetal liver cells (1×10^6) into tolerant, non-tolerant and control mice (Carrier et al. 1995). One mouse that was durably engrafted in the peripheral blood and the liver and permanently tolerant by skin graft acceptance showed a 5-fold and 25-fold increase in donor cells in the peripheral blood 3 and 6 months after the boost, respectively. There were two mice that had either a transient presence of donor cells in the peripheral blood with subsequent negative engraftment studies in the blood and tissues or no evidence of engraftment at any time as well as skin graft rejection. Neither mouse had evidence of circulating donor cells at 2 weeks, and 1, 2, 6 and 9 months following the postnatal boost. Five control, non-transplanted

BALB/c mice did not show any evidence of donor cells in the peripheral circulation at 2 weeks, and 1, 2, and 6 months post-transplant.

We wanted to test the hypothesis that although tolerance to donor skin was not induced in most engrafted animals, specific tolerance to donor HSC might be present. If this were true then the degree of engraftment (percentage of donor cells) should significantly increase with repeated injections of large numbers of donor cells (Stewart et al. 1993; Rao et al. 1997). We injected groups of allogeneic fetal liver and adult marrow recipients postnatally with large numbers (20–40×10^6 cells per day for 5 days) of donor cells. Female control (not transplanted) Balb/c mice were treated either with syngeneic or allogeneic mismatched (C57Bl/6) adult male marrow or pooled fetal liver cells. The syngeneic control recipients all engrafted, comparable with what has been previously reported in Balb/c mice (Stewart et al. 1993; Rao et al. 1997). The allogeneic mismatched naive recipients all rejected the donor cells by 4 weeks. A total of 6 fetal liver recipients and 7 adult BM recipients were studied in this experiment. In spite of repeated injections of large numbers of cells, we were unable to significantly increase the percentage of donor cells in these engrafted mice, supporting the idea that microchimerism in and of itself may not be sufficient to induce tolerance (George et al. 1998).

10.4.2.6 CD86-Deficient Dendritic Cell Progenitors

The results of our studies and those of others in animal models and humans indicate that microchimerism does not guarantee that tolerance will be induced and that even animals with up to 2–3% donor cells could still reject donor skin grafts. Thus, we developed the hypothesis that induction of tolerance was the first critical step in a successful in utero transplant. While competition with host HSC remains a major obstacle, it appeared to us that one way to initiate the process of significant durable engraftment was to induce tolerance to donor HSC. This would allow a post-transplant approach that might include repeated injections of donor cells with or without conditioning to reduce the host stem cell compartment. We thus began experiments to evaluate the effects of co-injecting immature donor dendritic cells along with HSC on engraftment and tolerance induction. We hypothesized that mature donor DC in the preparations might activate donor T cells to "make space" in the recipient and that immature DC (lacking the second signal) would tolerize the recipient immune system to donor MHC.

DC are considered very potent antigen-presenting cells, but evidence also exists that immature DC are strongly tolerogenic (Thomson et al. 1995). For example, co-stimulatory molecule (CD86)-deficient donor DC progenitors can prolong adult cardiac allograft survival in non-immunosuppressed murine recipients. Using our MHC-mismatched allogeneic mouse model, we examined the effect of donor DC progenitors on the degree of engraftment following in utero transplantation. BALB/C (GPI-1^a, H2^d, Thy1.2) were recipients, and male C57BL/6 (GPI-1^b, H2^b, Thy1.2) were donors, as previously described (Carrier et al. 1995).

We generated DC precursors from male C57BL/6 mice (H2K^b, I-A^b) by culturing BM in the presence of GM-CSF for 10 days. The method was modified after that described by Inaba et al. (1992) and Lu et al. (1995). Adult (10–12 weeks) male C57Bl/6 mice were used as the source of BM cells. For immature DC, 2×10^6 BM cells were cultured for 6 days in 24 well plates in medium supplemented with 10 ng/ml rm GM-CSF. For mature DC, BM cells were cultured for 10 days in medium supplemented with 6 ng/ml rm IL-4 in addition to rm GM-CSF. At the end of the culture period, the non-adherent mononuclear cells were collected. To enrich for DC, granulocytes were removed using biotin-conjugated anti-mouse granulocyte monoclonal antibody and magnetic bead-conjugated streptavidin (Dynal, NY). The enriched DC preparation was characterized by flow cytometry. The phenotype of the resulting cell preparations was 60% CD11c$^+$, 90% CD45$^+$, 50% CD80$^+$, 4% CD86$^+$ and <1%CD3$^+$.

Balb/c (H2K^d, I-A^d) fetuses were injected IP. The procedure for in utero transplantation was a modification of our previously described method (Carrier et al. 1995, 1996). Briefly, 15- to 16-day-old pregnant BALB/c mice were anesthetized with an isoflurane. Fetal recipients were transplanted IP using a 45–50 μm glass micropipette. Each injection consisted of 1.5×10^6 viable fresh adult donor BM cells either with or without 2×10^5 enriched donor DC in 5 μl of PBSC. An additional control group received 2×10^5 enriched donor DC alone. Peripheral blood was obtained at regular intervals beginning at 1 month of age. Recipient mice were anesthetized with isofluorine anesthesia and killed by cervical dislocation at 9 months of age. Blood and various tissues

Table 3. Engraftment of donor cells in the blood of recipients of DC progenitors in utero

	n	Engraftment	H2K^{b} [1]	Poly/Mono[2]	Lymphs[2]	CD3^{+} Lymphs[3]
DC	16	4/16 (25%)	0.82±0.21	0.04±.002	0.04±.004	0.29±.01
BM	19	12/19 (64%)	1.01±0.08	0.12±.02	0.04±.003	0.03±.002
BM+DC	19	7/19 (37%)	40.6±1.8	12.5±4.4	12.9±4.7	7.7±2.5

[1] Percentage of cells expressing H2K^{b}.
[2] Percentage of cells in the polymorphonuclear/monocyte gate or lymphocyte gate by FACS.
[3] Percentage of lymphocytes expressing CD3.

(see below) were collected for engraftment and histologic evaluation. Engraftment was determined by flow cytometry for H2K^{b}-positive cells and quantitative PCR for H2K^{b} (in males) or the Y chromosome (in females) in the peripheral blood of recipients up to 8 months of age. The percentage of donor cells was evaluated by fluorescent cell surface staining with monoclonal antibodies and analysis on a FACScan (Becton Dickinson, San Jose, CA). The FITC-, Biotin- and Tricolor- were conjugated to rat or mouse monoclonal antibodies to mouse macrophages (F4/80), granulocytes, T cells (CD3e, CD4, CD8), B cells (B220), leukocytes (CD45), NK cells (pan-DX5), and to donor cells (H2K^{b}).

The results (mean±SE) of FACS analysis at 1 month of age are shown in Table 3. These results were confirmed by the quantitative PCR (data not shown). While the mice that received both BM and DC had a comparable engraftment rate at 1 month of age, more importantly, the degree of engraftment was significantly greater than that in either the BM or DC recipients alone. By 8 months of age, none of the recipients of either DC or BM alone had detectable donor cells in the peripheral blood, while 2/15 recipients of DC+BM remained positive for circulating donor cells, with 93% of the cells being of donor origin.

Interestingly, four recipient mice in the BM+DC group died at about 1 month of age with weight loss and histologic evidence of GVHD. The two long-term survivors in that group also had evidence of GVHD at 8 months of age. None of the animals in the other two groups had any evidence of GVHD.

To evaluate tolerance induction, we placed donor skin grafts on a subgroup of animals at 6 months of age in each treatment group. Skin grafting was modified from our previously described method (Carrier et al. 1995, 1996). Briefly, syngeneic and allogeneic (C57BL/6) donor ear skin was used. The ear graft was placed on the graft bed so the graft did not overlap with the surrounding normal skin, and a bandage was placed over the graft. After 9 days, the bandage was removed and grafts were scored daily thereafter. Animals that rejected donor skin grafts within 14 days of placement were classified as non-tolerant. The only animals to durably accept donor skin grafts were the two long-term strongly engrafted recipients of DC+BM. These results suggest that in the presence of donor DC, donor T cells can be activated resulting in GVHD and increased donor cell engraftment. However, DC progenitors (CD86⁻) are not sufficient by themselves to induce tolerance to donor MHC. Whether this is a result of the presence of CD80 expression and/or the rapid maturation of these cells in vivo to CD86⁺ remains to be determined.

10.5 Summary and Conclusions

The potential advantage of in utero HSC transplantation over a postnatal BMT is that early curative therapy could be given to an affected fetus, thus eliminating standard intensive immunosuppressive, marrow-ablative conditioning. It is apparent from studies in animals and humans that MHC-mismatched donor HSC of either fetal or adult origin can engraft in fetal recipients if the transplants are done sufficiently early in gestation. However, except for SCID, the percentage of donor pluripotent HSC that engraft is unacceptably low. We had hoped that for diseases such as thalassemia there would be a selective survival advantage for committed donor progenitor cells resulting in a high percentage of donor cell engraftment. At least based upon the experience in human fetuses with α- or β-thalassemia, this has not been the case. Furthermore, for the majority of potential recipients of in utero HSC transplants, the marrow is non-defective, and the small percentage of pluripotent donor HSC that engraft would not be expected to selectively expand post-transplant.

Our own results suggest that the non-defective fetal mouse and rhesus monkey are excellent models in which to study both stem cell engraftment, rejection, and tolerance induction. In our studies in nondefective mice with normal hematopoiesis, while the percentage of donor cells that are present is quite low, in only a small number of these animals were we able to induce permanent skin graft tolerance. Thus, while we found microchimerism in ~75% of recipients, less than 10% became tolerant. Even when we co-injected a large number of DC precursors, similar to what has been shown to induce tolerance to allogeneic liver, most of the animals failed to become tolerant to donor skin grafts. Interestingly, donor c-kit$^+$ cells can be recruited with cytokines into the peripheral blood in engrafted mice, although these cells do not seem to be sufficient to induce tolerance to donor skin grafts, suggesting that the type (and location) of the engrafted donor cell plays a key role in tolerance induction. Our results in the fetal monkey model parallel those in the mouse, i.e., only a small number of donor cells engraft with limited tolerance induction. Interestingly, we found in our study of DC that GVHD was induced in those murine recipients of both allogeneic marrow and DC. It is likely that there were a sufficient number of mature DC in the preparation to facilitate a donor cytotoxic response towards the host. As a consequence there was also a significant increase in the percentage of donor cells that engrafted in the survivors. Future studies will focus on ways of blocking the graft vs host reaction while still maintaining the graft-promoting role of the donor T cell.

Acknowledgements. Supported by grants from the National Institutes of Health NIAID PO1-AI29512 and NHLBI RO1-HL58842, and the March of Dimes (FY95–0954).

References

Archer D, Turner C, Yeager A, Fleming W (1997) Sustained multilineage engraftment of allogeneic hematopoietic stem cells in NOD/SCID mice after in utero transplantation. Blood 90:3222

Aversa F, Tabilio A, Terenzi A et al (1994) Successful engraftment of T-cell-depleted haploidentical "three-loci" incompatible transplants in leukemia patients by addition of recombinant human granulocyte colony-stimulating factor-mobilized peripheral blood progenitor cells to bone marrow inoculum. Blood 84:3948–3955

Bennett SRM, Carbone FR, Karamalis F et al (1998) Help for cytotoxic T cell responses is mediated by CD40 signalling. Nature 393:478–480

Bensinger W, Weaver C, Appelbaum FR et al (1995) Transplantation of allogeneic peripheral blood stem cells mobilized by recombinant human granulocyte colony-stimulating factor. Blood 85:1655

Blazar B, Taylor P, Linsley P, Vallera D (1994) In vivo blockage of CD28/CTLA4:By/BB1 interaction with CTLA4-Ig reduces lethal murine graft-versus-host disease across the major histocompatibility complex barrier in mice. Blood 83:3815

Blazar B, Taylor P, McElmurry R, Tian L Panoskaltsis-Mortari A, Lam S, Lees C, Waldschmidt T, Vallera D (1998) Engraftment of severe combined immune deficient mice receiving allogeneic bone marrow via in utero or postnatal transfer. Blood 92:3949

Bodine DM, Seidel NE, Orlic D (1996) Bone marrow collected 14 days after in vivo administration of granulocyte colony-stimulating factor and stem cell factor to mice has 10-fold more repopulating ability than untreated bone marrow. Blood 88:89

Brady K, Cowan M, Leavitt A (1996) Circulating re cells usually remain of host origin after bone marrow transplantation for severe combined immunodeficiency. Transfusion 36:314–317

Carrier E, Lee TH, Busch M, Cowan MJ (1995) Induction of tolerance after in utero transplantation of major histocompatibility complex mismatched fetal stem cells in nondefective mice. Blood 86:4681–4690

Carrier E, Lee TH, Busch M, Cowan MJ (1996) Recruitment of engrafted donor cells into the peripheral circulation with Stem Cell Factor and Granulocyte-Colony Stimulating Factor following in utero transplantation of MHC-mismatched fetal liver and bone marrow cells in a nondefective murine model. Transplantation 64:1–7

Caux C, Massacrier C, Vanbervliet B et al (1994) Activation of human dendritic cells through CD40 crosslinking. J Exp Med 180: 1263–1272

Christensen R (1989) Hematopoiesis in the fetus and neonate. Pediatr Res 26:531–535

Cowan M, Wara D, Weintrub P, Ammann A (1985) Haploidentical bone marrow transplantation for severe combined immunodeficiency disease using soybean agglutinin-negative, T depleted marrow cells. J Clin Immunol 5:370–376

Cowan M, Tarantal A, Capper J, Harrison M, Garovoy M (1996) Longterm engraftment of T cell depleted parental marrow transplanted into fetal rhesus monkeys. Bone Marrow Transplant 17:1157–1165

Cowan MJ (1991) Bone marrow transplantation for the treatment of genetic diseases. Clin Biochem 24:1

Cowan MJ, Golbus M (1994) In utero hematopoietic stem cell transplants for inherited diseases. Am J Pediatr Hem Oncol 16:35–42

Crispe IN (1994) Fatal interactions: Fas-induced apoptosis of mature T cells. Immunity 1:347–349

Crombleholme TM, Harrison MR, Zanjani E (1990) In utero transplantation of hematopoietic stem cells in sheep: the role of T cells in engraftment and graft-versus host disease. J Pediatr Surg 25:885

de La Selle V, Gluckman E, Bruley-Rosset M (1996) Newborn blood can engraft adult mice without inducing graft versus host disease across non H-2 antigens. Blood 87:3977

de St Groth BF (1998) The evolution of self-tolerance: a new cell arises to meet the challenge of self-reactivity. Immunol Today 19:448–454

Dror Y, Gallagher R, Wara D, Weintrub P, Cowan M (1993) Immune reconstitution in severe combined immunodeficiency disease after lectin-treated, T cell depleted haplocompatible bone marrow transplantation. Blood 81:2021–2030

Flake A, Roncarolo M-G, Puck J, Almeida-Porada G, Evans M, Johnson M, Abella E, Harrison D, Zanjani E (1996) Treatment of X-linked severe combined immunodeficiency by in utero transplantation of paternal bone marrow. N Engl J Med 335:1806

Flake A, Zanjani E (1993) In utero transplantation of hematopoietic stem cells. Crit Rev Oncol Hematol 15:35

Flake A, Zanjani E (1999) In utero hematopoietic stem cell transplantation: ontogenic opportunities and biologic barriers. Blood 94:2179–2191

Fleischman R, Mintz B (1979) Prevention of genetic anemias in mice by microinjection of normal hematopoietic cells into the fetal placenta. Proc Natl Acad Sci USA 76:5736

Fleischman R, Mintz B (1984) Development of adult bone marrow stem cells in H-2-compatible and -incompatible mouse fetuses. J Exp Med 159:731–745

Freeman G, Gribben J, Boussiotis V, Ng J, Restivo V, Lombard L, Gray G, Nadler L (1993) Cloning of B7-2: A CTLA-4 counter-receptor that costimulates human T cell proliferation. Science 262:909

George J, Sweeney S, Kirklin J, Simpson E, Goldstein D, Thomas J (1998) An essential role for Fas ligand in transplantation tolerance induced by donor bone marrow. Nat Med 4:333–335

Gil J, Porta F, Bartolome J, Lafranchi R et al (1999) Immune reconstitution after in utero bone marrow transplantation in a fetus with severe combined immunodeficiency with natural killer cells. Transplant Proc 31:2581

Green J, Noel P, Sperling A, Walunas T, Gray G, Bluestone J, Thompson C (1994) Absence of B7-dependent responses in CD28 deficient mice. Immunity 1:501

Hahn T, Shulman L, Ben-Hur H et al (1994) Differential responses of fetal, neonatal, and adult myelopoietic progenitors to interferon and tumor necrosis factor. Exp Hematol 22:114

Hajdu K, Tanigawara S, McLean L, Cowan MJ, Golbus M (1996) In utero allogeneic hematopoietic stem cell transplantation to induce tolerance. Fetal Diagn Ther 11:241

Harding F, McArthur J, Gross J, Raulet D, Allison J (1992) CD28-mediated signalling co-stimulates murine T cells and prevents induction of anergy in T cell clones. Nature 356: 607

Harrison M, Slotnick RN, Crombleholme TM et al (1989) In utero transplantation of fetal liver haemopoietic stem cells in monkeys. Lancet ii:1425

Hayward A, Ambruso D, Battaglia F, Donlon T, Eddelman K, Giller R, Hobbins J, Hsia YE, Quinones R, Shpall E, Trachtenberg E, Giardina P (1998) Microchimerism and tolerance following intrauterine transplantation and transfusion for a-thalassemia. Fetal Diagn Ther 13:8

Hobbs JR (1988) Displacement bone marrow transplantation and immunoprophylaxis for genetic diseases. Adv Intern Med 33:81

Honey K, Cobbold S, Waldmann H (1999) CD40 ligand blockage induces CD4+ T cell tolerance and linked suppression. J Immunol 163:4805–4810

Inaba K, Steinman R, Pack M, Aya H, Inaba M, Sudo T, Worpe S, Schuler G (1992) Identification of proliferating dendritic cell precursors in mouse blood. J Exp Med 175:1157–1167

Jones DRE, Bui TH, Anderson EM et al (1996) In utero haematopoietic stem cell transplantation: current perspectives and future potential. Bone Marrow Transplant 18:831–837

Jordan C, McKearn J, Lemischka R (1990) Cellular and developmental properties of fetal hematopoietic stem cells. Cell 61:953

Karmann K, Hughes CC, Schechner AB et al (1995) CD40 on human endothelial cells: inducibility by cytokines and functional regulation of adhesion molecule expression. Proc Natl Acad Sci USA 92:4342–4346

Kelemen E, Calvo W, Fliedner T (1979) Atlas of Human Hemopoietic Development. Springer-Verlag, New York

Kirk A, Burkly L, Batty D, Baumgartner R et al (1999) Treatment with humanized monoclonal antibody against CD154 prevents acute renal allograft rejection in nonhuman primates. Nat Med 5:686–693

Kruisbeek A, Amsen D (1996) Mechanisms underlying T-cell tolerance. Curr Opin Immunol 8:233–244

Larsen CP, Elwood ET, Alexander DZ et al (1996) Long-term acceptance of skin and cardiac allografts after blocking CD40 and CD28 pathways. Nature 381:434–438

Lenschow D, Zeng Y, Thistlethwaite J, Montag A, Brady W, Gibson M, Linsley P, Bluestone J (1992) Long-term survival of xenogeneic pancreatic islet grafts induced by CTLA4Ig. Science 257:789

Lin H, Bolling S, Linsley P, Wei R, Gordon D, Thompson C, Turka L (1993) Long-term acceptance of major histocompatibility complex mismatched cardiac allografts induced by CTLA4-Ig plus donor-specific transfusion. J Exp Med 178:1801

Lu L, McCaslin D, Starzl T, Thomson A (1995a) Mouse bone marrow-derived dendritic cell progenitors (NLDC 145$^+$, MHC class II$^+$, B7–1^{dim},B7–2$^-$) induce alloantigen-specific hyporesponsiveness in murine T lymphocytes. Transplantation 60:1539

Lu L, Rudert W, Qian S et al (1995b) Growth of donor-derived dendritic cells from the bone marrow of murine liver allograft recipients in response to GM colony-stimulating factor. J Exp Med 182:379

Martin P (1996) Prevention of allogeneic marrow graft rejection by donor T cells that do not recognize recipient alloantigens: potential role of a veto mechanism. Blood 88:962

Merino A, Dror Y, Benkerrou M, Cowan M (1993) Development of tolerance after haplocompatible T-depleted bone marrow transplantation. Bone Marrow Transplant 12:481–488

Muench M, Cupp J, Polakoff J, Roncarolo M (1994) Expression of CD33, CD38, and HLA-DR on CD34$^+$ human fetal liver progenitors with a high proliferative potential. Blood 83:3170

Mychaliska G, Rice H, Tarantal A, Stock P, Capper J, Garovoy M, Olson J, Cowan M J, Harrison M (1997) In utero hematopoietic stem cell transplants prolong survival of postnatal kidney transplants in monkeys. J Pediatr Surg 32:976–981

Ogawa M, Nishikawa S, Yoshinaga K et al (1993) Expression and function of c-kit in fetal hemopoietic progenitor cells: transition from the early c-kit-independent to the late c-kit-dependent wave of hemopoiesis in the murine embryo. Development 117:1089

O'Reilly R, Brochstein J, Dinsmore R, Kirkpatrick D (1984) Marrow transplantation for congenital disorders. Semin Hematol 21:118

Papayannopoulou T, Brice M, Broudy V, Zsebo K (1993) Isolation of c-kit receptor-expressing cells from bone marrow, peripheral blood and fetal liver: Functional properties and composite antigenic profile. Blood 78:1403

Qin S, Cobbold SP, Pope H et al (1993) "Infectious" transplantation tolerance. Science 259:974

Ramsdell F, Fowlkes BJ (1990) Clonal deletion versus clonal anergy: the role of the thymus in inducing self tolerance. Science 248: 1342

Ramshaw H, Crittenden R, Dooner M, Peters S, Rao S, Quesenberry P (1995) Biology of blood and marrow. Transplantation 1:74–80

Rao S, Peters S, Crittenden R, Stewart F, Ramshaw H, Quesenberry P (1997) Stem cell transplantation in the normal nonmyeloablated host: relationship between cell dose, schedule and engraftment. Exp Hematol 25:114

Reitsma M, Harrison M, Pallavicini M (1993) Detection of a male-specific sequence in nonhuman primates through use of the polymerase chain reaction. Cytogenet Cell Genet 64:213–216

Ridge J, Di Rosa F, Matzinger P (1998) A conditioned dendritic cell can be a temporal bridge between a CD4+ T-helper and a T-killer cell. Nature 393:474–478

Roodman GD, Kuehl TJ, Vandeberg JL, Muirhead DY (1991) In utero bone marrow transplantation of fetal baboons with mismatched adult baboon marrow. Blood Cells 17:376

Rossini A, Greiner D, Mordes J (1999) Induction of immunologic tolerance for transplantation. Physiol Rev 79:99–141

Sayegh M, Turka L (1998) The role of T-cell costimulatory activation pathways in transplant rejection. N Engl J Med 338:1813–1821

Sayegh M, Zheng X-G, Magee C, Hancock W, Turka L (1997) Donor antigen is necessary for the prevention of chronic rejection in CTLA4IG-treated murine cardiac allograft recipients. Transplantation 64:1646–1650

Schoenberger SP, Toes RE, Van der Voort EI et al (1998) T cell help for cytotoxic T lymphocytes is mediated by CD40-CD40L interactions. Nature 393:440–443

Sharma S, Karak A, Sharma D, Aggarwal A (1991) Hemopoietic stem-cell differentiation in fetal liver, spleen, and thymus of rhesus monkeys. Med Oncol Tumor Pharmacother 8:113–4

Slaper-Cortenbach I, Ploemacher R, Löwenberg B (1987) Different stimulative effects of human bone marrow and fetal liver stromal cells on erythropoiesis in long-term culture. Blood 69:135

Srour EF, Zanjani ED, Brandt JE et al (1992) Sustained human hematopoiesis in sheep transplanted in utero during early gestation with fractionated adult human bone marrow cells. Blood 79:1404

Stewart F, Crittenden R, Lowry P et al (1993) Long-term engraftment of normal and post-5-fluorouracil murine marrow into normal nonmyeloablated mice. Blood 81:2566

Stiehm ER, Roberts R, Hanley-Lopez J, Wakim ME, Pallavicini M, Cowan MJ, Ettenger RB, Feig SA (1996) Successful bone marrow transplantation in severe combined immunodeficiency from a sibling that received a paternal bone marrow transplant. N Engl J Med 335:1811–1814

Tarantal A (1990) Interventional ultrasound in pregnant macaques: embryonic/fetal applications. J Med Primatol 19:47

Tarantal A (1993) Hematologic reference values for the fetal long-tailed macaque (Macaca fascicularis). Am J Primatol 29:209

Tarantal A, Cowan M (1999) Administration of stem cell factor (SCF) and granulocyte-colony stimulating factor (G-CSF) to maternal and fetal rhesus monkeys. Cytokines 11:290–300

Tarantal A, Hendrickx AG (1988) Prenatal growth in the cynomolgus and rhesus macaque (Macaca fascicularis and Macaca mulatta): a comparison by ultrasonography. Am J Primatol 15:309

Tarantal A, Goldstein O, Barley F, Cowan M (2000) Transplantation of human peripheral blood stem cells (PBSC) into fetal rhesus monkeys (*Macaca mulatta*). Transplantation 69:1818–1823

Thomas J, Carver F, Kasten-JollyJ et al (1994) Further studies of veto activity in rhesus monkey bone marrow in relation to allograft tolerance and chimerism. Transplantation 57:101

Thomson A, Lu L, Murase N et al (1995) Microchimerism, dendritic cell progenitors and transplantation tolerance. Stem Cells 13:622

Tocci A, Rezzoug F, Wahbi K, Touraine J-L (1995) Fetal liver generates low CD4 hematopoietic cells in murine stromal cultures. Blood 85:1463

Touraine J (1991) In utero transplantation of fetal liver stem cells in humans. Blood Cells 17:379

Touraine J (1996) Treatment of human fetuses and induction of immunological tolerance in humans by in utero transplantation of stem cells into fetal recipients. Acta Haematol 96:115

Van Den Heuvel R, Schoeters G, Leppens H, Vandervorght O (1991) Stromal cells in long-term cultures of liver, spleen, and bone marrow at different developmental ages have different capacities to maintain GM-CFC proliferation. Exp Hematol 19:115

Wagnum A, Westerman T, Visser T, Wagemaker G (1994) Distribution of receptors for GM-CSF on immature $CD34^+$ bone marrow cells, differentiating mono myeloid progenitors and mature blood cell subsets. Blood 84:764

Walunas T, Lenschow D, Bakker C, Linsley P, Freeman G, Green J, Thompson C, Bluestone J (1994) CTLA-4 can function as a negative regulator of T cell activation. Immunity 1:405

Wengler G, Lanfranchi A, Frusca T, Verardi R et al (1996) In utero transplantation of parental CD34 haematopoietic progenitor cells in a patient with X-linked severe combined immunodeficiency. Lancet 348:1484–1487

Westgren M, Ringden O, Eik-Nes S et al (1996) Lack of evidence of permanent engraftment after in utero fetal stem cell transplantation in congenital hemoglobinopathies. Transplantation 61:1176

Yamamoto Y, Yasumizu R, Amou Y et al (1996) Characterization of peripheral blood stem cells in mice. Blood 88:445

Zheng L, Fisher G, Miller R, Peschon J, Lynch D, Lenardo M (1995) Induction of apoptosis in mature T cells by tumor necrosis factor. Nature 377:348–352

11 Fetal Hematopoiesis During First and Second Trimester of Pregnancy: Relevance for In Utero Stem Cell Transplantation

J. Wilpshaar, H. H. H. Kanhai, S. Scherjon, J. H. F. Falkenburg

11.1 Introduction

In-utero stem cell transplantation (IUST) is an intriguing alternative to postnatal transplantation of hematopoietic stem cells. However, despite decades of research, both in animal models and in human fetuses, the clinical application of IUST is still not feasible. Thus far, more than 30 case reports on clinical experience with IUST in the human fetus, for various indications, have been published or described (Table 1). These cases were treated with cells from fetuses of various gestational ages and with different quantity and quality of stem cells. Only a few cases of IUST in severe combined immunodeficiency (SCID) showed some de-

Table 1. Overview of clinical experience with HSC-IUTP in the human

Reference	Diagnosis	GA (weeks)	Transplant	Number	iv/ip	Engraftment
1. Linch et al. 1986	Rh(D) immunization	17	Maternal BM, Campath	15×10^7 NC; 5.4×10^5 HPC	iv	No
2. Touraine et al. 1988	BLS	28	Fetal liver and thymus (2 fetuses: 7 and 7.5 weeks)	16×10^6	iv	10–26% of lymphocytes
3. Roncarolo et al. 1989	SCID	26	Fetal liver and thymus (1 fetus: 7.5 weeks)	37×10^6 fetal liver cells	iv	Engraftment of T-lymphocytes
4. Touraine 1990	β-Thalassemia	12	Fetal liver (fetus 9.5 weeks)	3×10^8	ip	0.9–30% HbA
5. Touraine 1990	CGD	18	Fetal liver		iv	Bradycardia, death
6. Touraine 1991; Touraine et al. 1991a–c, 1992; Raudrant et al. 1992	β-Thalassemia	17	Fetal liver (11.5 weeks)		iv	Bradycardia, death
7. Slavin et al. 1990, 1992	GLD	34	Paternal BM, T-cell depleted (Campath)	33×10^8	ip	No
8. Slavin et al. 1990, 1992	GLD	23	Paternal BM, T-cell depleted (Campath)	30×10^8	intra-portal and ip	No
9. Slavin et al. 1990, 1992	β-Thalassemia	25	HLA-matched sibling BM,T-cell depleted (Campath)	30×10^8	intra-portal and ip	
10. Diukman and Golbus 1992	α-Thalassemia	18	Maternal BM, T-cell depleted		ip	Term 24 weeks, autopsy: micr engraft
11. Diukman and Golbus 1992	SCID	20	Maternal BM, T-cell depleted		ip and iv	Term 26 weeks, autopsy: no engraftment
12. Diukman and Golbus 1992	Chediak-Hiagashi	19	Maternal BM, T-cell depleted		ip	No
13. Thiliganthan et al. 1993	Rh(D) antagonism	12	Maternal BM, T-cell depleted	23×10^6 MNC	ip	No
14. Cowan and Golbus 1994	SCID	20	Maternal BM, T-cell depleted		iv, ip comb.	No donor cells at 24 weeks, terminated
15. Cowan and Golbus 1994	β-Thalassemia	14	Fetal liver			No, septic abortion
16. Flake et al. 1995, 1996	SCID	16, 17.5, 18.5	Paternal BM, CD34 enriched, T-cell depleted	14.8×10^6; 2×10^6; 1.8×10^6	ip	100% engraftment of donor T-lymphocytes
17. 1995 Flake	CGD	131415	Paternal BM, CD34 enriched		ip	No

Table 1. Continued

Reference	Diagnosis	GA (weeks)	Transplant	Number	iv/ip	Engraftment
18. Westgren et al. 1996	α-Thalassemia	15, 31	Fetal livers (5, 6, 7, 9, 9, 10, 10, 8, 9, 10)	219×10^6; 270×10^6	ip, iv	No
19. Westgren et al. 1996	Sickle cell anemia	13	Fetal livers (6, 8, 8, 10, 10)	82×10^6	ip	No
20. Westgren et al. 1996	β-Thalassemia	18	Fetal livers (6, 7, 8, 9, 10)	188×10^6	iv	No
21. Wengler et al. 1996	SCID	21, 22	Paternal BM, CD34 enriched			T cells in CB
22. Hayward et al. 1996	α-Thalassemia	131823	Paternal BM, CD34 enriched			Trace
23. Touraine 1993; Touraine et al. 1993	Hurler disease	14	Fetal liver			Low level
24. Touraine 1996a,b	Niemann-Pick	14	Fetal liver			Yes
25. Blakemore et al. 1996	GLD	13	Paternal BM, CD34 enriched	5×10E9 per kg	ip	Fetus died at 20 weeks
26. Blakemore et al. 1996	GLD	13	Paternal BM, CD34 enriched	5×10E9 per kg	ip	No, or little
27. Blakemore et al. 1996	GLD	13	Paternal BM, CD34 enriched	5×10E8 per kg	ip	No, or little
28. Porta et al. 1998	SCID X1	21,22	Paternal BM, CD34 enriched, T-cell depleted	14×10E6; 4. 8×10E6	ip	Engraftment of T and NK cells
29. Porta et al. 1998	SCID (T-, B-, nk+)	22,23	Paternal BM, CD34 enriched, T-cell depleted	24×10E6; 10×10E6	ip	None at birth; 2 months 6% CD3+
30. Porta et al. 1998	SCID (T-, B-, nk+)	22,23	Paternal BM, CD34 enriched, T-cell depleted	20×10E6; 3×10E6 stromal cells	ip	T cells

BLS, bare lymphocyte syndrome; *CGD*, chronic granulomatous disease; *GLD*, globoid cell dystrophy/metachromatic leukodystrophy; *SCID*, severe combined immunodeficiency disease; *SCID Xl*, X-linked SCID; *iv*, intravenous; *ip*, intraperitoneal, HbA, hemoglobine A; *GA*, gestational age

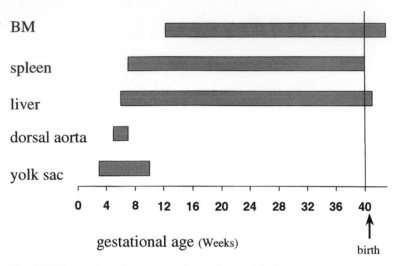

Fig. 1. Different sites where stem cells are detected during pregnancy

gree of engraftment. A potential cause of failure of IUST is lack of engraftment due to competition of the transplant with a vigorous hematopoietic compartment in the host. Not only differences in the characteristics of stem cells used can influence engraftment, but also characteristics of stem cells present in the host at each specific gestational age have to be taken into account. During embryonic and fetal life, the major hematopoietic sites of hematopoiesis change (Fig. 1). We have summarized the literature on the hematopoietic compartment of the first and second trimester fetus and discussed it in relation to IUST.

11.2 Yolk Sac

Hematopoiesis in the human yolk sac occurs as early as day 18 of development (week 4^{+4} of gestation; Bloom and Bartelmez 1940; Tavian et al. 1999). It has long been assumed that yolk sac-derived precursors are also responsible for the colonization and, hence, hematopoietic development of the other blood-forming tissues that sequentially developed in the embryo and fetus, i.e., liver, thymus, spleen and eventually, bone marrow (BM) (Metcalf and Moore 1971; Andrew and Owen 1978). More recently, it was suggested that hematopoietic stem cells (HSC) are generated from the same ancestor cell early in development, which gave rise to several different extra-embryonic (yolk sac) and embryonic tissues (embryonic mesoderm; Rich 1995; Van Zant et al. 1997). In the extra-embryonic area, both vascular and hematopoietic systems originate as a network of "hemangioblastic" mesodermal cell aggregates. The most peripheral cells in these originally solid clusters flatten into endothelial cells, while most of the inner ones simultaneously disappear and thus form the first vessel lumens. Blood islands are formed, and here hematopoiesis begins (Sabin 1920; Kessel and Fabian 1985). Hematopoiesis in the yolk sac is restricted to erythropoiesis. Moreover, there is no nuclear expulsion (Marks and Rifkind 1972), and nucleated red cells are therefore produced, i.e., as an endproduct of yolk sac erythropoiesis.

11.3 Circulating Blood

The first hematopoietic $CD45^+CD34^-$ cells have been detected inside the embryo at day 22 of development (5 weeks of gestation; Tavian et al. 1996), just following the onset of cardiac contractions. Macrophages are probably the first blood cells to appear and neutrophils the last (Christensen 1989). Forestier observed in blood from 50 fetuses at 18–22 weeks gestation, 80±9% lymphocytes, 5±2% neutrophils, 1.5±2% monocytes, and 12±8% normoblasts (Forestier et al. 1986). In contrast to the blood of adults, fetal blood contained high concentrations of hematopoietic progenitors. The percentage of $CD34^+$ cells in fetal blood at 17–24 weeks of gestation was higher compared with that at-term gestation (37–41 weeks of gestation), 6.4±1.3% to 1.3±0.2% respectively, which was a 4.9-fold decrease (Shields and Andrews 1998). Significant changes in the frequency of $CD34^+$ cells did not occur until after 28 weeks of gestation. Also, the percentage of $CD34^+CD38^-$ cells was significantly higher in second trimester blood compared with term blood (0.72±0.26% vs 0.06±0.02% respectively; Subek et al. 1998).

11.4 Dorsal Aorta

Tavian et al. (1996) identified in the pre-umbilical region of the 5-week human embryo a large population of hematopoietic cells associated with the ventral endothelium of the dorsal aorta. These cells exhibited a surface phenotype ($CD45^+CD34^+Lin^-$), in vitro behavior and a gene expression pattern that were characteristic of primitive hematopoietic progenitors (Labastie et al. 1998). Furthermore, the culture of these dissected hematopoietic aortic clusters on a stromal cell line which supported hematopoiesis generated much higher numbers of $CD34^+$ cells and progenitors than any other part of the embryo, including the liver (Tavian et al. 1999). In addition, these aortic cells have greater reproductive capability than HSC obtained from umbilical cord blood (UCB) and adult BM (Auerbach et al. 1996). In both bird and mouse embryos, also has been demonstrated that the aortic region is a primary hematopoietic territory (Dieterlen-Lievre and Le Douarin 1993; Dieterlen-Lievre 1997).

11.5 Liver

It is from the sixth week of gestation that hematopoiesis is found in the fetal liver (Paul et al. 1969; Migliaccio et al. 1986). Initiation of hepatic hematopoiesis is heralded by the appearance of undifferentiated blast cells in the hepatic cords. Differentiation is then gradually seen not only in erythroid lineage, but also in granulocytic and megakaryocytic lines (Barker et al. 1969). However, the liver is predominantly erythropoietic. Hematopoietic progenitor cells (HPC), that are apparent in human fetal liver (FL) until 8-weeks of gestation, differ in several aspects from adult BM-derived progenitors. These differences include: their very rapid cycling rate (Peschle et al. 1981), the constant and rapid expanding pool size (Migliccio et al. 1986), in vivo differentiation almost exclusively along the erythroid pathway (Kelemen et al. 1979), and accelerated in vitro maturation time (Peschle et al. 1981).

Roy et al. (1997) showed that for adult BM, FL colony-forming cells (CFCs) were $CD34^+$/HLA-DR$^+$/CD38$^+$. Adult BM long-term culture-initiating cells (LTC-ICs) were $CD34^{++}$/CD38$^-$/HLA-DR$^{-/dim}$, and FL LTC-ICs were $CD34^{+++}$/CD38$^-$, which in addition also expressed high levels of HLA-DR antigens. An explanation for the presence of HLA-DR-positive antigens on FL LTC-ICs is the higher proliferative state of FL LTC-ICs.

It has been suggested that in the early second trimester of gestation, the fetus might be immunologically immature. However, the FL already contained 5×10^6 T cells at 16 weeks of gestation, which increased to 100×10^6 at 22 weeks. Not only FL, but also other fetal organs from 14 weeks of gestation onwards, proved to contain normal-to-high frequencies of alloreactive cytotoxic T-lymphocyte precursors (CTLP) and helper T-lymphocyte precursors (HTLP; Lim et al. 1996).

11.6 Spleen

Historically, the human fetal spleen has been regarded as an organ of hematopoiesis (Heller et al. 1947; Linman and Bethell 1957; Wolf et al. 1983). At about 7 weeks of gestation, the first hematopoietic cells are found in the fetal spleen. Between 14 and 22 weeks of gestation, $4.2 \pm 2.0\%$ of the nucleated cells are CD34$^+$ (Lim et al. 1996). The precise role of the spleen during normal hematopoietic development does not appear to be as clear as is sometimes suggested. The spleen can become a site of hematopoiesis during myeloproliferative disorders or infection (Ward and Block 1971; Stallmach and Karolyi 1994). Recently it has been suggested that the mid-gestation spleen does not normally serve as a significant granulocytopoietic or erythropoietic organ (Calhoun et al. 1996). The precursor cells found in the spleen might be a result of entrapment of the hematopoietic cells from the fetal blood, which contains numerous HSC (Wolf et al. 1983). Similar percentages of lymphocytes, neutrophils and normoblasts observed for fetal blood and spleen support this. The spleen ceases to contain hematopoietic cells at birth, but could become hematopoietic in abnormal situations.

11.7 Bone Marrow

The last hematopoietic station present during fetal development is the BM. BM stroma first appear in the clavicle at 9–10 weeks of gestation.

Active hematopoiesis follows about 3 weeks later at 12–13 weeks of gestation (Kelemen et al. 1979). In the onset of BM hematopoiesis, five stages were identified by Charbord et al. (1996). From 6.6 to 8.5 weeks of gestation, only entirely cartilaginous rudiments were found. At stage 2 (8.5–9 weeks) chondrolysis was actively proceeding. The development of BM is subsequent to the penetration of avascular cartilage by perichondral mesenchymal cells and their associated blood vessels. This leads to calcification of cartilage (Petrakis et al. 1969). Development of the vascular bed occurred between 9 and 10.5 weeks, but hematopoiesis was not detected until the 4th stage, from 10.5 –15 weeks of gestation. The final stage (16 weeks onwards) was characterized by the final organization of the long bones, with areas of fully calcified bone and areas of dense hematopoiesis. At 22 weeks of gestation all cell lines are represented in the BM (Chervenick 1981). Absolute numbers of CD34+ cells increase exponentially from 21×10^6 at 17–18 weeks to 224×10^6 at 21–22 weeks of gestation (Wilpshaar et al. 1998).

11.8 Comment

For years, it has been assumed that hematopoiesis originated in the yolk sac and that from here stem cells migrated to the fetal liver and spleen and finally colonized the BM (Johnson and Moore 1974). It was suggested that migration from one organ to the other might occur via the bloodstream (Metcalf and Moore 1971; Migioccio et al. 1986; Tavassoli 1994). The yolk sac/FL pathway was subject to query after the identification of an intra-embryonic pool of HSC in the pre-umbilical region of the 5-week human embryo associated with the ventral endothelium of the dorsal aorta (Tavian et al. 1996). When HSC from this region were transplanted in SCID mice, hematopoietic reconstitution was obtained (Godin et al. 1993). The significance of these findings was that HSC are found transiently in this region, before the initiation of fetal liver hematopoiesis, and therefore represented the first intra-embryonic site of hematopoiesis. Since blood already circulates when HSC appear in the ventral wall of the dorsal aorta, interchanges of precursors between embryo and yolk sac may have already occurred, thus complicating the interpretation of the origin of aorta-associated hematopoietic cells.

Recent studies suggested a new theory, in which primitive (embryonic) and definitive (fetal/adult) HSC are generated from the same ancestor cells early during development. This common ancestor, the primordial germ cell (PGC), initially resides inside the embryo and gives rise to several different extra-embryonic and embryonic tissues, a process which is called PGC allocation (Lawson and Hage 1994; Rich 1995; Van Zant et al. 1997). To become germ cells, PCGs must migrate from the extra-embryonic sites back into the embryo, thereby passing through the region that comprises the ventral wall of the dorsal aorta. Tavian et al. (1996) described the first intra-embryonic HSC population in this area as being a transient population even before hepatic hematopoiesis occurred. It is still unclear whether these PGC-allocated HSC remain in each organ and migrate to the next hematopoietic site or whether they are continuously in circulation and seed the different sites depending upon the time in ontogeny. The latter hypothesis suggested that restricted numbers of PGC-allocated HSC are present in each organ, and that the quantity and/or quality of these cells determined how long hematopoiesis will be maintained in those respective organs during ontogeny (Van Zant et al. 1997).

During the second trimester, HSC are found at different places at the same time during ontogeny. The BM contributes to the stem cell pool from the beginning of the second trimester of pregnancy, which is earlier than was previously thought. Therefore, the division into the different hematopoietic compartments was not correct. It seems more likely that there is a pool of stem cells that divides itself over different compartments. This seeding of stem cells does not exclude the possibility that stem cells might leave that organ again; however, with regard to the expanding numbers, most of them are more likely to stay. Furthermore, the environment of the organs might influence the percentage of stem cells present. Homing might be different at specific times during ontogeny. Age- and tissue related differences in the biological characteristics of progenitor cells are large and therefore seem important for the success of IUST. The study of its biology may provide important clues as to how the progenitor pool is maintained.

The success of IUST is also source-dependent. Recent developments showed that in an NOD/SCID model, fetal cells engrafted independently of their phase in cell cycle where adult cells only engrafted in quiescent state (Wilpshaar et al. 1999). Therefore, fetal cells might be a favorable source for in utero transplants. The only successful in utero transplantations which have taken place in SCID patients suggest that immunological issues play an important role.

Acknowledgements. The authors gratefully acknowledge Dr. G.C. Beverstock for critical review of the manuscript.

References

Andrew TA, Owen JJ (1978) Studies on the earliest sites of B cell differentiation in the mouse embryo. Dev Comp Immunol 2:339–346

Auerbach R, Huang H, Lu L (1996) Haematopoietic stem cells in the mouse embryonic yolk sac. Cells 14:269–280

Barker JE, Keenan MA, Raphals L (1969) Development of the mouse hematopoietic system. II. Estimation of spleen and liver "stem" cell number. J Cell Physiol 73:51–56

Blakemore K, Bambach B, Moser H, Corson V, Griffin C, Noga S, Periman E, Wenger D, Zuckerman R, Khouzami A, Jones R (1996) Engraftment following in utero bone marrow transplantation for globoid cell leukodystrophy. Am J Obstet Gynecol 312 (abstract)

Bloom W, Bartelmez GW (1940) Hematopoiesis in young human embryos. Am J Anat 67:21–53

Calhoun DA, Li Y, Braylan RC, Christensen RD (1996) Assessment of the contribution of the spleen to granulocytopoiesis and erythropoiesis of the mid-gestation human fetus. Early Hum Dev 46:217–227

Charbord P, Tavian M, Humeau L, Peault B (1996) Early ontogeny of the human marrow from long bones: an immunohistochemical study of hematopoiesis and its microenvironment. Blood 87:4109–4119

Chervenick (1981) Atlas of blood cells

Christensen RD (1989) Hematopoiesis in the fetus and neonate (review) (88 refs). Pediatr Res 26:531–535

Cowan MJ, Golbus M (1994) In utero hematopoietic stem cell transplants for inherited diseases. Am J Pediatr Hematol Oncol 16:35–42

Dieterlen-Lievre F, Le Douarin NM (1993) Developmental rules in the hemopoietic and immune systems of birds: how general are they? Semin Dev Biol 4:325–332

Dieterlen-Lievre F (1997) Intraembryonic hematopoietic cells. In: Zon L (ed) Hematology/oncology clinics of North America. Saunders, Philadelphia, pp 1149–1171

Diukman R, Golbus MS (1992) In utero stem cell therapy. J Reprod Med 37:515–520

Flake AW, Puck JM, Almeida-Porada G, Evans MI, Johnson MP, Roncarolo MG, Zanjani ED (1995) Successful treatment of X-linked recessive severe combined immunodeficiency (X-SCID) by the in utero transplantation of CD34 enriched paternal bone marrow cells. Blood 86:486

Flake AW, Roncarolo MG, Puck JM, Almeida-Porada G, Evans MI, Johnson, MP, Abella EM, Harrison DD, Zanjani ED (1996) Treatment of X-linked severe combined immunodeficiency by in utero transplantation of paternal bone marrow (see comments). N Engl J Med 335:1806–1810

Forestier F, Daffos F, Galacteros F, Bardakjian J, Rainaut M, Beuzard Y (1986) Hematological values of 163 normal fetuses between 18 and 30 weeks of gestation. Pediatr Res 20:342–346

Godin I, Garcia-Porrero JA, Coutinho A, Dieterlen-Lievre F, Marcos MAR (1993) Para-aortic splancnopleura from early mouse embryos contains B 1 a cell progenitors. Nature 364:67–70

Hayward A, Hobbins J, Quinones RR et al (1996) Microchimerism and tolerance following intrauterine transplantation and transfusion for a-thalassemia 1. Conference proceedings. In utero stem cell transplantation and gene therapy, Reno Nevada, 1996

Heller EL, Lewishon MG, Palin WE (1947) Aleukemic myelosis, chronic non-leukemic myelosis, agnogenic myeloid metaplasia, osteosclerosis, leukoerythroblastic anemia, and synonymous designations. Am J Pathol 23:327

Johnson GR, Moore MAS (1974) Role of stem cell migration in initiation of mouse foetal liver haemopoiesis. Nature 258:726

Kelemen E, Calvo W, Fliedner (1979) Intravascular hemopoietic cells. In: Kelemen E, Calvo W, Fliedner (eds) Atlas of human hematopietic development. Springer, Berlin Heidelberg New York, p 51

Kessel J, Fabian BC (1985) Graded morphogeneic patterns during the development of the extraembryonic blood system and coelom of the chick blastoderm: a scanning electron microsope and light microscope study. Am J Anat 173:99–112

Labastie MC, Cortes F, Romeo PH, Dulac C, Peault B (1998) Molecular identity of hematopoietic precursor cells emerging in the human embryo. Blood 92:3635

Lawson KA, Hage WJ (1994) Clonal analysis of the origin of primordial germ cells in the mouse. In: Germline development. Wiley, Chichester, p 68

Lim FTH, van Luxemburg-Heijs SAP, Kanhai HHH, Beekhuizen W, Willemze R, Falkenburg JHF (1996) Normal fetal T cell allo-reactivity in early second trimester: relevance for in utero stem cell transplantation. Blood 88:597

Linch DC, Rodeck CH, Nicolaides K, Jones HM, Brent L (1986) Attempted bone-marrow transplantation in a 17-week fetus (letter). Lancet 2:1453

Linman JW, Bethell FH (1957) Agnogenic myeloid metaplasia: its natural history and present day management. Am J Med 22:107

Marks PA, Rifkind RA (1972) Protein synthesis: its control in erythropoiesis. Science 175:955–961

Metcalf D, Moore MAS (1971) Embryonic aspects of haemopoiesis. In: Neuberger A, Tatum EL (eds) Haemopoietic cells. North Holland, Amsterdam,pp 172–266 (Frontiers in biology, vol 24)

Migliaccio G, Migliaccio AR, Petti S, Mavilio F, Russo G, Lazzaro D, Testa U, Marinucci M, Peschle C (1986) Human embryonic hemopoiesis: kinetics of progenitors and precursors underlying the yolk sac-liver transition. J Clin Invest 78:51–60

Paul J, Conkie D, Freshney RI (1969) Erythropoietic cell population changes during the hepatic phase of erythropoiesis in the foetal mouse. Cell Tissue Kinet 2:283–294

Peschle C, Migliaccio AR, Migliacco G, Ciccariello R, Lettieri F, Quattrin S, Russo G, Mastroberardino G (1981) Identification and characterization of three classes of erythroid progenitors in human fetal liver. Blood 58:565–572

Petrakis NL, Pons S, Lee RE (1969) An experimental analysis of factors affecting the localization of embryonic bone marrow. In Vitro 4:3–13

Porta, Lanfranchi A, Tettoni K, Verardi R, Mazzolari E, Zucca A, Carlo-Stella C, Notarangelo LD, Ugazio AG (1998) Stromal and CD34 positive cells IUT in SCID. 3rd International symposium on in utero stem cell transplantation and gene therapy

Raudrant D, Touraine JL, Rebaud A (1992) In utero transplantation of stem cells in humans: technical aspects and clinical experience during pregnancy. Bone Marrow Transplant 9 [Suppl 1]:98–100

Rich IN (1995) Primordial germ cells are capable of producing cells of the hematopoietic system in vitro. Blood 86:463

Roncarolo MG, Bacchetta R, Bigler M, Touraine JL, de Vries JE, Spits H (1991) A SCID patient reconstituted with HLA-incompatible fetal stem cells as a model for studying transplantation tolerance. Blood Cells 17:391–402

Roy V, Miller JS, Verfaillie CM (1997) Phenotypic and functional characterization of committed and primitive myeloid and lymphoid hematopoietic precursors in human fetal liver. Exp Hematol 25:387–394

Sabin FR (1920) Studies on the origin of blood vessels and of red blood corpuscles as seen in the living blastoderm of chicks during the second day of incubation. Contrib Embryol 9:214–262 (Carnegie Inst Wash Pub no 272)

Shields LE, Andrews RG (1998) Gestational age changes in circulating CD34[+] hematopoietic stem/progenitor cells in fetal cord blood. Am J Obstet Gynecol 178:931–937

Slavin S, Naparstek E, Ziegler M, Bach G, Schenker JG, Lewin A (1990) Intrauterine bone marrow transplantation for correction of genetic disorders in man. Exp Hematol 18:658

Slavin S, Naparstek E, Ziegler M, Lewin A (1992) Clinical application of intrauterine bone marrow transplantation for treatment of genetic diseases – feasibility studies. Bone Marrow Transplant 9:189–190

Stallmach T, Karolyi L (1994) Augmentation of fetal granulopoiesis with chorioamnionitis during the second trimester of gestation. Hum Pathol 25:244–247

Surbek DV, Holzgreve W, Jansen W, Heim D, Garritsen H, Nissen C, Wodnar-Filipowicz A (1998) Quantitative immunophenotypic characterization, cryopreservation, and enrichment of second- and third-trimester human fetal cord blood hematopoietic stem cells (progenitor cells). Am J Obstet Gynecol 179:1228–1233

Tavassoli M (1994) Embryonic origin of hematopoietic stem cells (editorial; comment). Exp Hematol 22:7–7

Tavian M, Coulombel L, Luton D, Clemente HS, Dieterlen-Lievre F, Peault B (1996) Aorta-associated CD34+ hematopoietic cells in the early human embryo. Blood 87:67–72

Tavian, Hallais MF, Peault B (1999) Emergence of intraembryonic hematopoietic precursors in pre-liver human embryo. Development 126:793–803

Thilaganthan B, Nicolaides KH, Morgan G (1993) Intrauterine bone-marrow transplantation at 12 weeks' gestation (letter). Lancet 342:243–243

Touraine JL (1990) In utero transplantation of stem cells in humans. Nouv Rev Fr Hematol 32:441–444

Touraine JL (1991) In utero transplantation of fetal liver stem cells in humans. Blood Cells 17:379–387

Touraine JL (1993) Transplantation of fetal liver stem cells into patients and into human fetuses, with induction of immunologic tolerance. Transplant Proc 25:1012–1013

Touraine JL (1996a) In utero transplantation of fetal liver stem cells into human fetuses (review). J Hematother 5:195–199

Touraine JL (1996b) Treatment of human fetuses and induction of immunological tolerance in humans by in utero transplantation of stem cells into fetal recipients. Acta Haematol 96:115–119

Touraine JL, Raudrant D, Royo C, Rebaud A, Roncarolo MG, Souillet G, Philippe N, Touraine F, Betuel H (1989) In-utero transplantation of stem cells in bare lymphocyte syndrome (letter). Lancet 1:1382–1382

Touraine JL, Raudrant D, Royo C, Rebaud A, Barbier F, Roncarolo MG, Touraine F, Laplace S, Gebuhrer L, Betuel H et al (1991a) In utero transplantation of hemopoietic stem cells in humans. Transplant Proc 23:1706–1708

Touraine JL, Laplace S, Rezzoug F, Sanhadji K, Veyron P, Royo C, Maire I, Zabot MT, Vanier MT, Rolland MO et al (1991b) The place of fetal liver transplantation in the treatment of inborn errors of metabolism. J Inherit Metab Dis 14:619–626

Touraine JL, Raudrant D, Vullo C, Frappaz D, Freycon F, Rebaud A, Barbier F, Roncarolo MG, Gebuhrer L, Betuel H et al (1991c) New developments in stem cell transplantation with special reference to the first in utero transplants in humans (review) (9 refs). Bone Marrow Transplant 7 [Suppl 3]:92–97

Touraine JL, Raudrant D, Rebaud A, Roncarolo MG, Laplace S, Gebuhrer L, Betuel H, Frappaz D, Freycon F, Zabot MT et al (1992) In utero transplantation of stem cells in humans: immunological aspects and clinical follow-up of patients. Bone Marrow Transplant 9 [Suppl 1]:121–126

Touraine JL, Roncarolo MG, Bacchetta R, Raudrant D, Rebaud A, Laplace S, Cesbron P, Gebuhrer L, Zabot MT, Touraine F et al (1993) Fetal liver transplantation: biology and clinical results (review) (5 refs). Bone Marrow Transplant 11 [Suppl 1]:119–122

Van Zant G, de Haan G, Rich IN (1997) Alternatives to stem cell renewal from a developmental viewpoint. Exp Hematol 25:187

Ward H, Block M (1971) The natural history of agnogenic myeloid metaplasia (AMM) and a critical evaluation of its relationship with the myeloproliferative syndromes. Medicine 50:357–420

Wengler GS, Lanfranchi A, Frusca T, Verardi R, Neva A, Brugnoni D, Giliani S, Fiorini M, Mella P, Guandalini F, Mazzolari E, Pecorelli, S, Notarangelo LD, Porta F, Ugazio AG (1996) In-utero transplantation of parental CD34 haematopoietic progenitor cells in a patient with X-linked severe combined immunodeficiency (SCIDXI). Lancet 348:1484–1487

Westgren M, Ringden O, Eik-Nes S, Ek S, Anvret M, Brubakk AM, Bui TH, Giambona A, Kiserud T, Kjaeldgaard A, Maggio A, Markling L, Seiger A, Orlandi F (1996) Lack of evidence of permanent engraftment after in utero fetal stem cell transplantation in congenital hemoglobinopathies. Transplantation 61:1176–1179

Wilpshaar J, Joekes EC, Lim FTH, Kanhai HHH, Willemze R, Bloem JL, Falkenburg JHF (1998) Magnetic resonance imaging to characterize the human hematopoietic stem cell compartment during fetal development. Blood 92:268

Wilpshaar J, Falkenburg JHF, Kanhai HHH, Breese R, Srour EF (1999) Engraftment potential of human fetal liver and bone marrow CD34+ cells residing in different phases of cell cycle. Blood 94:34

Wolf BC, Luevano E, Neiman RS (1983) Evidence to suggest that the human fetal spleen is not a hematopoietic organ. Am J Clin Pathol 80:140

12 Tolerance Induction Following In Utero Stem Cell Transplantation

D.R.E. Jones, E.M. Anderson, D.T.Y. Liu, R. M. Walker

12.1 Introduction

One of the central tenets underpinning the science of immunology is the concept that the immune system is able to recognise "self" tissues and consequently, to distinguish "self" from "non-self". This ability to discriminate, even at a subtle molecular level, confers powerful protective functions on the immune system because invasion from without, by potentially harmful pathogenic agents, can be repulsed very effectively by various interacting arms of this host defence system. Dogma asserts that the tolerance to self tissues is learned in utero, during embryonic/fetal development (Billingham et al. 1953). Thus, as the immune system itself develops, the repertoire essential to effect host defence in postnatal life – the ability to discriminate between self and non-self – becomes indelible. However, there is a period in human fetal development when the immune system can be considered to be naive, that is, without full competence to discriminate it is unable to mount an immune response and would therefore leave the developing fetus vulnerable to invasion from without. This period, essentially during the first trimester of (human) fetal life, is predominantly prior to the phase at which passive immunity can be conferred via transfer of maternal immunoglobulins. Under normal circumstances, the uterine environment protects the fetus from infection, and development of immunocompetence can, therefore, proceed without impediment.

This period of immunological naivety provides the ideal opportunity for transplantation: a recipient lacking the immunological apparatus to mount a response to foreign tissue. This "window of opportunity" is relatively short, and by around 13–15 weeks' gestation, lymphoid cells can be detected within the human fetal liver (Jones et al. 1995) with implications for development of a capacity to recognise foreign tissue. This gestational stage is also significant in haematopoietic development. Prior to 13–15 weeks' gestation, haematopoiesis proceeds within the human fetal liver, but after this time, active migration of cells occurs to seed the bone marrow (BM), spleen and lymphoid system. It is during these events that immunological competence is developing and lymphocytes appear within the fetal liver (Jones et al. 1995), but it is a matter for debate whether these cells can be regarded as immunologically functional at this gestational age (Ek et al. 1994). Distinct subsets of T cells are found within the liver in adult life, and these cells are considered not to have been "educated" via migration through the thymus (Norris et al. 1998). Such cells might have a different role in immunological development (Norris et al. 1998), so the presence of lymphocytes in the fetal liver need not be indicative of the capacity to recognise and reject foreign tissue.

To date, there have been around 32 human cases of in utero stem cell transplantation performed worldwide, but not all have been successful. Paradoxically (in the light of our knowledge of immunological development), it is in utero transplantation for immunodeficiencies which have had the most successful outcomes. This situation has prompted dissenters to conclude that it is the presence of a functional immune system which precludes engraftment in those disorders where the transplant has apparently failed to provide clinical improvement. In order to improve the utility of in utero transplantation as a viable therapeutic option for genetic disorders which can be diagnosed early in prenatal life, a number of questions about fetal haematopoietic development require to be addressed, and our definition of immunological tolerance may need reappraisal.

12.2 Animal Models

In order to study mechanisms of engraftment after in utero stem cell transplantation (and to devise means to improve the success of the procedure), it has been necessary to resort to animal models which allow the creation of a chimaera: i.e. containing both human and indigenous cells. The mouse has been a very useful model for this purpose, but in most strains a compliant genetic background has been engineered to facilitate sustained human donor cell engraftment. In this respect, the NOD/severe combined immunodeficiency syndrome (SCID) mouse is ideal, and engraftment with human haematopoietic cells (including lymphoid cells) can be achieved (Cashman et al. 1997). However, this model does not extrapolate readily to human: there is little development of the immune repertoire in utero, and there is no hepatic, pre-immune phase of haematopoiesis. Alternatively, the sheep has been established as a large animal model for in utero transplantation (Zanjani et al. 1991): it requires no conditioning, it has a relatively long gestation period (approximately 145 days) and undergoes a pre-immune, hepatic, haematopoietic developmental phase during fetal development. The sheep model of in utero transplantation can be used to address questions regarding levels of engraftment with haematopoietic stem cells (HSC) obtained from various sources, optimal numbers and timing of transplants, and whether tolerance to donor cells is required for sustained levels of engraftment to be achieved.

12.2.1 Results: Sheep Transplants

To date, at this centre, we have performed in utero transplantation in 44 sheep fetuses, at 55 ± 2 days' gestational age. For this study, human fetal haematopoietic cells which had been stored (liquid nitrogen) (Jones et al. 1995) were used in the transplant procedure. Dorset-cross ewes were sedated, placed in dorsal recumbency, and under general anaesthesia laparotomy was performed to deliver the uterus on to the abdominal wall. The fetus was palpated through the uterine wall and stabilised by hand whilst the cell suspension was injected into the fetal abdominal cavity. The injected preparations (each 0.5 ml) were either human fetal haematopoietic cell suspensions (gestational age: 6–17 weeks, $n=36$) or

Table 1. Outcome of in utero transplantation (IUT) procedures in 44 fetal sheep recipients. The data are tabulated according to the source of (human) donor cells used. All transplants were performed at 55±2 days' gestation

Source of donor cells[a]	Number of recipients engrafted
FL6 (n=5)	3
FL8 (n=11)	9[b]
FL9 (n=6)	4
FL10 (n=5)	1
FL12 (n=6)	2
FL17 (n=3)	0
BM MNC (n=4)	0
Control[c] (n=4)	(0)

FL, fetal liver-derived haematopoietic cells, followed by gestational age in weeks; *BM MNC*, (adult) bone marrow mononuclear cell preparation

[a] n, number of fetal sheep recipients of each donor cell type.

[b] Includes 3 fetal sheep sacrificed at IUT + 22 days.

[c] Control, RPMI only injected in utero.

mononuclear cell preparations derived from (normal, healthy, adult) human BM donations (n=4). As a control, 0.5 ml of tissue culture medium alone, (n=4), was injected. For each transplant, the "loaded" syringes were kept on ice and then warmed in the hands immediately prior to injection. Finally, penicillin was injected into the amniotic cavity as a precaution, the uterus replaced and the abdominal wall closed. The animals were removed to a recovery room immediately post-surgery. The subsequent course of each pregnancy which continued to term was uneventful and 39 healthy lambs were born at 143±2 days. [The loss rate – 2 fetuses (4.5%) – is not regarded as unusual for husbandry conditions pertaining at this institution]. Three fetuses were killed 3 weeks post-transplant (day 77) and tissue samples taken for analysis to determine evidence of engraftment. With newborn lambs, blood and BM samples were first obtained at around 3 weeks (i.e. post-weaning). Evidence of engraftment with human cells was investigated using flow cytometry (human-specific antibody reagents) and by analysis of DNA prepared from the sheep BM mononuclear cells (Perkin-Elmer Quantiblot system). There was evidence of engraftment in 19 (41%) of the lambs using both DNA analysis and flow cytometry (Table 1). Levels of engraftment varied between 1% and 4% (using flow cytometry) and were consistently around 1–2% using DNA analysis.

The data obtained from these studies indicate that the number of engrafted recipients is greatest when haematopoietic cells derived from human fetal liver, obtained during the first trimester of gestation, are used as the transplant material. However, a negative result, as assessed using these techniques, need not necessarily be indicative of a lack of engraftment. The sensitivity of the Quantiblot method is around 1%, and it is difficult to provide an objective assessment for the sensitivity of the flow cytometric analysis, given the relatively low numbers of human cells being sought within the recipient tissue. Thus, unless the levels of engraftment in the sheep recipient are relatively high (perhaps 1% or above), it is difficult to be certain that donor cells are indeed present. Certainly, our data indicate that sustained levels of engraftment can be achieved when haematopoietic cells obtained from first trimester (<12 weeks) human fetal liver are used as the transplant and that the degree of engraftment decreases when fetal liver from a later gestational period provides the donor cells.

12.2.2 Results: Sources of Donor Cells

Fetal liver-derived HSC obtained between 6 weeks and around 10 weeks' gestational age produce predominantly erythroid progeny in culture, in vitro (Jones et al. 2000). This apparent commitment to the erythroid lineage begins to diminish at around 11 weeks, and by 15 weeks it is no longer apparent (Jones et al. 2000). We have used this information to select samples of fetal HSC, with diminishing erythroid commitment, to use in in utero transplantation (IUT) procedures in the sheep model. When donor cells for IUT were obtained from the early gestational period (<10 weeks), cells showing the same erythroid commitment were found in the recipient sheep BM up to 2 years postnatally. For analysis of the presence of engraftment, antibodies to glycophorin A (Immunotech) and to CD34 (HPCA-2: Becton-Dickinson) were used, neither of which show evidence of cross-reactivity with sheep cells. To demonstrate multilineage engraftment, an antibody to CD45 (BRA-55: Sigma), which also does not cross-react with sheep haematopoietic cells, was used. Human CD45-positive cells could not be detected, in

BM, in those animals which had received the early gestation (<10 weeks) fetal haematopoietic cells in utero. To further determine the erythroid commitment of these transplanted cells, we investigated responses, within the recipient sheep haematopoietic system, to injections of human G-CSF. A 3-day injection regimen resulted in an increase in the number of glycophorin-A-positive cells in the BM and an increase in CD34-positive cells (Jones et al. 1997). Again, no CD45-positive cells were detected. However, it was noted that the response to human G-CSF in the recipient sheep was relatively short lived, possibly due to specific antibody production negating the effects of G-CSF. However, these data are an indication that active human haematopoiesis can be retained within the recipient (sheep) haematopoietic tissue, which is capable of expanding in response to human haematopoietic growth factors.

When later gestation fetal cells (17 weeks) were transplanted in utero, very little evidence of engraftment was observed in the recipients postnatally, using the same analytical methodology as that which detected engraftment after early gestation fetal haematopoietic cell transplants. Low levels of CD45-positive cells were detected within the recipient BM (<1%), but these cells were not present on every occasion when biopsies were taken. Where BM-derived mononuclear cell preparations had been used as the transplant material, no human cells could be detected in recipients. However, it is uncertain whether this result reflects a lack of engraftment, or engraftment at a level which is below that which can be detected with the methods used. Recently, we conducted studies to follow the routes of engraftment with fetal liver-derived haematopoietic cells, after IUT. It is apparent that the liver is the initial site of homing for the donor cells and that higher levels of engraftment (12–18%) are detected at this site, within 3 weeks, post-transplant, when early gestation human fetal liver-derived haematopoietic cells are used (manuscript in preparation).

It is noteworthy that, in all instances, the levels of human cells detected within sheep recipients fluctuated over the periods of time (up to 3 years in some cases) during which analyses were performed. It is possible that this phenomenon is due to variations inherent in analytical methods, although a real variation in numbers of cells present at any given time cannot be ruled out.

12.3 Conclusions

Available data show that T cells are not produced within the first trimester in human fetal liver (Jones et al. 1995; Jaleco et al. 1997) and that thymic education is necessary for functional T cell development to occur (Plum et al. 1994; Bodey et al. 1998). Thus, during the period of fetal immunological (T cell) naivety, rejection of foreign tissue is unlikely. We have studied this phenomenon in the sheep by in utero transplantation, in early gestation, of human haematopoietic cells and the finding of long-term engraftment in this model is strongly indicative of an immunological microenvironment receptive to even xenogeneic donor cells. Thus, we surmise that tolerance to these foreign cells has occurred, facilitated by the immunological naivety of the fetal recipient. Experiments of nature have served to underscore these findings: the presence of blood group chimerism in animals laid the foundations for our current thinking on immunological tolerance (Owen 1945), and more recently it has been shown that relatively high proportions of human twins are chimeric for their sibling's blood group (van Dijk et al. 1996). Such data suggest that tolerance to major tissue antigen mismatches is achievable and that it could have a significant clinical impact if it were to be induced by in utero transplantation of appropriate cells. In this context, perhaps it is more relevant to consider other findings from the twin chimerism study (van Dijk et al. 1996): in a number of cases, although having evidence of a major ABO incompatibility, there were normal titres of anti-A or anti-B, but the cells were not destroyed by the offending (i.e. incompatible) antibody.

Attempts to generate animal models of in utero tolerance, although having mixed success, have nevertheless provided encouraging data for discussion. Studies designed to induce tolerance in utero by transplantation of HSC and thereafter to seek evidence of enhanced engraftment by postnatal boosting with cells of the same origin, show that it is indeed possible to achieve tolerance to the donor cell type in a mouse model (Carrier et al. 1995). Although the level of engraftment achieved by IUT was less than 1%, postnatal boosting produced an increase in the numbers of circulating donor cells over and above the numbers transplanted de novo (Carrier et al. 1995). The indication that the presence of a low number of engrafted cells in a recipient animal are sufficient for the induction and maintenance of tolerance is in general agreement with data obtained from our sheep studies. Here we showed that (a) sustained engraftment can be achieved after transplant of human fetal haematopoietic cells and (b) these cells respond to human haematopoietic growth factors. Thus, as in the mouse study (Carrier et al. 1995), we conclude that the haematopoietic microenvironment must be tolerant to the mismatched tissue. Other workers have demonstrated that, in animals defined as tolerant in this way after IUT, there is no in vitro response between recipient and donor cells in cytotoxicity or in mixed lymphocyte reactions (Cowan et al. 1996) further underlining the immunological basis for lack of rejection. Similar results have been obtained in a human patient following an IUT procedure to correct α-thalassaemia (Hayward et al. 1998) and serve to emphasise that immune rejection of the graft is unlikely to be the reason for the limited donor cell engraftment observed in that case.

A natural progression from tolerance to haematopoietic cells by IUT would be to seek evidence of tolerance to other tissues as a result of this procedure. Such studies have been performed and have produced intriguing results. One group showed that, in a rhesus monkey model, prior in utero stem cell transplantation could at least partially tolerise recipients for a postnatal kidney graft from the same donor (Mychaliska et al. 1997). In a mouse model, it has been shown that using fetal liver-derived cells as the in utero transplant vehicle significantly increases the frequency of tolerance to other tissues grafted postnatally (Yuh et al. 1996), a phenomenon also demonstrated after injection of such cells into newborn mice with subsequent cardiac allografting (West et al. 1994).

Thus the potential to utilise the immunological immaturity of a fetus to facilitate a transplant in postnatal life is slowly being realised. Susceptibility to tolerance is greatest during immune development, which in humans and other animals with a long gestation, occurs in utero (West et al. 1994). The outcome of an IUT procedure should be considered to be both haematological chimerism and tolerance to the donor tissue type. Lack of clinically significant levels of engraftment after IUT could be viewed as an opportunity to intervene postnatally with a further transplant, if there are strong indications for a lack of reactivity to the donor tissue type. However, this situation does not preclude IUT for de facto correction of a disorder diagnosed in prenatal life. A number of haematological disorders are associated with irreversible pathological manifestations in utero, and there is an urgent requirement for therapeutic intervention early in gestation. Clearly, animal models will serve to address the problems which have been encountered in the clinic, in particular the poor success rate for IUT in conditions other than immunodeficiencies. The sheep model of IUT, with a long gestation and no requirement for conditioning to achieve engraftment, is a valuable addition to the tools available in this quest.

Acknowledgements. The authors are grateful to the Trent Regional Health Authority and, latterly, to the EU Biomed 2.0 programme for funding these studies.

References

Billingham RE, Brent L, Medawar PB (1953) 'Actively acquired tolerance' of foreign cells. Nature 172:603–606

Bodey B, Bodey B Jr, Siegel SE, Kaiser HE (1998) Intrathymic non-lymphatic hematopoiesis during mammalian ontogenesis. In Vivo 12:599–618

Carrier E, Lee TH, Busch MP, Cowan MJ (1995) Induction of tolerance in nondefective mice after in-utero transplantation of major histocompatibility complex-mismatched fetal hematopoietic stem-cells. Blood 86:4681–4690

Cashman J, Bockhold K, Hogge DE, Eaves AC, Eaves CJ (1997) Sustained proliferation, multi-lineage differentiation and maintenance of primitive human haemopoietic cells in NOD/SCID mice transplanted with human cord blood. Br J Haematol 98:1026–1036

Cowan MJ, Tarantal AF, Capper J, Harrison M, Garovoy M (1996) Long-term engraftment following in utero T cell-depleted parental marrow transplantation into fetal rhesus monkeys. Bone Marrow Transplant 17:1157–1165

Ek S, Ringdén O, Markling L, Westgren M (1994) Immunological capacity of human fetal liver cells. Bone Marrow Transplant 14:9–14

Hayward A, Ambruso D, Battaglia F, Donlon T, Eddelman K, Giller R, Hobbins J, Hsia YE, Quinones R, Shpall E, Trachtenberg E, Giardina P (1998) Microchimerism and tolerance following intrauterine transplantation and transfusion for alpha-thalassemia-1. Fetal Diagn Ther 13:8–14

Jaleco AC, Blom B, Res P, Weijer K, Lanier LL, Phillips JH, Spits H (1997) Fetal liver contains committed NK progenitors, but is not a site for development of CD34+ cells into T cells. J Immunol 159:694–702

Jones DRE, Anderson EM, Evans AA, Liu DTY (1995) Long-term storage of human fetal hematopoietic progenitor cells and their subsequent reconstitution – implications for in utero transplantation. Bone Marrow Transplant 16:297–301

Jones DRE, Liu DTY, Anderson EM, Lamming GE (1997) Transplantation of human fetal liver-derived haematopoietic stem cells into sheep, in utero. In: Ringdén O, Hobbs JR, Steward CG (eds) Correction of genetic disorders by transplantation IV. COGENT Trust, London, pp 130–136

Jones DRE, El-Sammak M, Anderson EM, Khan Z, Simpson N (2000) Changes in epitope expression on CD34+ve cells during ontogeny of hae- matopoiesis in the human fetal liver may correlate with alteration in lineage commitment. Br J Haematol (submitted)

Mychaliska GB, Rice HE, Tarantal AF, Stock PG, Capper J, Garovoy MR, Ol- son JL, Cowan MJ, Harrison MR (1997) In utero hematopoietic stem cell transplants prolong survival of postnatal kidney transplantation in monkeys. J Pediatr Surg 32:976–981

Norris S, Collins C, Doherty DG, Smith F, McEntee G, Traynor O, Nolan N, Hegarty J, O'Farrelly C (1998) Resident human hepatic lymphocytes are phenotypically different from circulating lymphocytes. J Hepatol 28:84–90

Owen RD (1945) Immunogenetic consequences of vascular anastomoses be- tween bovine twins. Science 102:400–401

Plum J, De Smedt M, Defresne MP, Leclercq G, Vandekerckhove B (1994) Human CD34+ fetal liver stem cells differentiate to T cells in a mouse thy- mic microenvironment. Blood 84:1587–1593

van Dijk BA, Boomsma DI, De Man AJM (1996) Blood-group chimerism in human multiple births is not rare. Am J Med Genet 61:264–268

West LJ, Morris PJ, Wood KJ (1994) Fetal liver haematopoietic cells and toler- ance to organ allografts. Lancet 343:148–149

Yuh DD, Gandy KL, Hoyt G, Reitz BA, Robbins RC (1996) Tolerance to car- diac allografts induced in utero with fetal liver cells. Circulation 94 [Suppl 9]:11304–11307

Zanjani ED, Mackintosh FR, Harrison MR (1991) Hematopoietic chimerism in sheep and nonhuman primates by in utero transplantation of fetal he- matopoietic stem cells. Blood Cells 17:349–363

13 In Utero Stem Cell Transplantation in Humans

M. Westgren, L.E. Shields

13.1 Introduction

The prospect of in utero fetal transplantation represents a potential major step forward in the management of patients with congenital hematological, metabolic and immunological disorders. Successful allogeneic in utero transplantations (IUTs) of hematopoietic stem cells HSC) have been reported in several animal models. The first successful transplantation in man was reported by Touraine et al. in 1989 in a case of bare lymphocyte syndrome. Since then more than 35 IUTs have been published or reported at national or international meetings. The transplantations have been reported for a variety of indications, at various gestational ages, and with different sources of the transplant. All cases of IUT performed in fetuses with normal immunological function have failed, and unequivocal engraftment could so far only be demonstrated in fetuses suffering from severe immunodeficiency disorders. Most publications reviewing in utero stem cell transplantation have stated that the fetus is the ideal candidate for therapy due to its immature immune system, which should permit the transplantation of non-human leukocyte antigen (HLA) matched cells and allow the developing fetus to recognize the donor cells as "self" (Flake 1986; Crombleholme et al. 1990, 1991; Flake et al. 1991; Touraine et al. 1991; Zanjani et al. 1991; Zanjani 1992; Westgren and Ringden 1994; Jones et al. 1996; Flake and Zanjani 1997, 1999; Surbek et al. 1999). In addition, there should be abundant and readily available hematopoietic niches from donor cells to engraft secondary to fetal growth. However, when one considers that all attempted allogeneic transplantations in immunocompetent human fetuses have failed, then doubts have to be cast on these theories and the whole concept needs to be reappraised. The purpose of this chapter is to provide a critical review of the clinical progress in this field.

13.2 Fetal Immunology

Many investigators have described the early second trimester fetus as preimmune, i.e., unable to mount a successful immunological attack against donor cells (Harrison 1989; Crombleholme et al. 1990; Flake et al. 1991). In fact, this theory has been one of the fundamental foundations of IUT. It is clear from available published reports of in utero stem cell transplantation in humans that this theory can be questioned. The only clinically successful in utero transplants have been carried out in severely immunological compromised fetuses, suggesting that the fetal immune system probably plays a vital role in the success of in utero fetal transplantation.

While there is not a large body of literature relative to the ontogeny of the fetal immune system, there are enough data to suggest that the fetus, even very early in gestation, has the immune capacity to significantly impact the success of in utero stem cell transplantation. We feel that these factors must be taken into account when strategies are developed for in utero therapy, and potential approaches to these problems are discussed at the end of the chapter. Fetal hematological development is thought to begin in the fetal yolk sac, progress to the fetal liver and to a less extent the fetal spleen, and finally, in the late second and early third trimester, the primary hematopoietic organ is the fetal bone marrow (BM) (Tavassoli 1991). Fetal immunological cells (NK cells, T cells, and B cells) all have different developmental epics and timing of their presence in the fetal circulation. The fetal thymus originally develops from the third brachial cleft and pouch. A primordial thymus is present at approximately 7 weeks' gestation. The thymus is first colonized with cells from the fetal liver at 8.5–9.5 weeks' gestation, and shortly after this 20–50% of cells in the fetal thymus express the common T-cell surface phenotypes (CD7, CD2). Between 12–13 weeks, cells within the fetal liver and spleen express the T-cell receptor (TCR). By 16 weeks' gestation the fetal thymus has distinct cortical and medullary regions, suggesting functional maturity. The presence of functional T cells can also be demonstrated in vitro by their response in mixed lymphocyte culture (MLC). MLC is a method frequently used as a marker for allogeneic recognition and prediction of potential rejection during post-natal BM transplantation. Recently published work from our laboratory (Lindton et al., in press) and that of Stites et al. (1974) has shown that fetal liver cells clearly can respond to allogeneic stimulation by 12 weeks' gestation, and positive MLC results can be observed as early as 9.5 weeks. Thus, despite their early gestational age, the fetus develops the capacity for allogeneic recognition early in gestation. At this time, additional work is needed to fully elucidate the role of T-cell function and its relation to the in utero graft failure.

The ontogeny of fetal B-cell development begins slightly later in gestation. As with fetal T cells, development begins in the fetal liver. Cells possessing early B-cell antigen expression (CD19 and CD20) can be seen by immunofluorescent studies as early as 7–8 weeks' gestation (Gathings et al. 1977). Surface expression of immunoglobulin (Ig) M can be noted as early as 9–10 weeks. Cells in the fetal circulation express the common B-cell antigen (CD20) at 14–16 weeks' gestation, and secretion of IgM has been noted as early as 15 weeks, but does not reach normal postnatal levels until 1 year of age. Secretion of IgG is first noted at 20 weeks' gestation, with full levels not reached until approximately 5 years of age. Due to the relatively late appearance of humoral immunity, it is unlikely that B-cell function significantly affects the success of in utero stem cell transplants.

The third component of fetal immunity is through natural killer cells (NK cells). These cells share a number of antigenic and functional similarities to T cells, but do not rearrange or express T-cell antigen receptor genes, nor do they require the presence of MCH class I or class II antigens. They function through direct cytotoxicity to foreign antigens, usually viral, tumor cell, or trophoblastic cell stimulation as well as antibody-dependent cell cytotoxicity (ADCC). When expressed as a percentage of the total lymphocytes, the proportion of NK cells in the fetal circulation is quite high (29% at 13 weeks; Thilaganathan et al. 1993) and they are likely to play an important role in fetal immune defense early in pregnancy. NK cells are also thought to play an important role in graft failure in postnatal BM transplantation (Lanier 1995). Based on their high number, their early presence, and their lack of antibody-mediated toxicity, it is likely that they play a significant role in the success of in utero stem cell transplantation.

13.3 Methodological Factors that May Influence IUT

13.3.1 Hematopoietic Stem Cells

Stem cells are the earliest precursors of all lympho-hematopoietic cells and are best described by their function, the ability to sustain lympho-hematopoiesis and to generate mature progeny of all lympho-hematopoietic lineages (i.e., RBCs, WBCs, platelets, and tissue macrophages). A number of specific cell surface markers, specifically CD34 and subpopulations of the CD34+ cells (CD38-, HLA-DR, and recently KDR), have been identified as cells that function as definitive stem cells (Muench et al. 1994; Thilaganathan et al. 1994; Opie et al. 1998; Ziegler et al. 1999). HSC can be isolated from fetal liver, umbilical cord blood, and BM. Cells expressing the CD34 cell surface antigen represent approximately 1–5% of BM cells, 1–10% of mononuclear cells in term and preterm umbilical cord blood, and 3–4% of mononuclear cells isolated from fetal livers. Additional phenotypic and in vitro proliferative responses are present in these different stem cell sources with greater proliferation and a higher percentage of the mononuclear cell population seen in the fetal relative to the adult sources of stem cells. It is unclear at this time whether the differences between fetal and adult stem cells will result in differences in in utero engraftment or levels of chimerism.

13.3.2 Cell Dose

A variable known to be a determinant of successful engraftment in BM transplantation is the cell concentration obtained in receptive sites in the recipient, which in turn is dependent on the cell dose, mode of administration and the size of the recipient. Experience with BM transplantation suggests that $1\times10^{6-7}$ cells/kg will provide an adequate cell dose and a good chance for engraftment. From these data, most investigators have extrapolated into the in utero situation and used doses in the range 10^6-10^{10} cells/kg when attempting to perform fetal transplantations. Transplantation studies in both humans and in animal models (mice, sheep, and non-human primates) all suggest that the number of cells transplanted influences the rate of engraftment and level of chimerism (Rao et al. 1997; Blomberg et al. 1998; Shields and Andrews 1999). If IUT truly takes place under permissive conditions, then transplantation of allogeneic HSC should be similar to a competitive repopulating model. Under these conditions, donor cells and circulating autologous fetal cells are competing for engraftment into the same hematopoietic niches. The postnatal syngeneic mice, using competitive repopulating, provide useful data that are likely also applicable in the fetus transplantation. In this model, the level of chimerism increases as the number of cells transplanted increases, and the donor cells are given over a number of days and not as a single injection (Rao et al. 1997). Daily injection, 40×10^6 BM cells for 5 days (total cell dose 8×10^9/kg), results in chimerism averaging $39\pm6\%$, whereas when the same number of cells were given as a single injection, the level of chimerism was only 3–11%. Similar findings have been noted in both allogeneic and xenogeneic fetal sheep studies, where increased levels of chimerism are present after multiple injections of donor cells relative to the same number of cells given as a single injection (Zanjani et al. 1997).

In the fetus, hematopoietic niches should expand with fetal growth. Therefore, in theory donor cells should have readily available hematopoietic niches or sites to engraft into. However, the fetal environment may limit the ability of allogeneic donor cells to effectively compete with autologous fetal HSC. In early fetal life, the frequency of circulation of autologous HSC is high (approximately 10% mononuclear cells at 14–16 weeks; Thilaganathan et al. 1994; Shields and Andrews 1998). Therefore, donor HSC without a selective advantage, such as those observed in immune deficiencies, may be at a competitive disadvantage relative to the high frequency of circulating endogenous HSC. These data suggest that large numbers of donor cells are likely to be needed to effectively compete with endogenous cells. In the xenogeneic and allogeneic fetal sheep model, the level of chimerism increased with increasing numbers of donor cells and peaked after a single cell dose of 3×10^8 cell/kg (Zanjani et al. 1997). However, when the same total number of donor cells was given in multiple doses (three), the level of chimerism increased to levels similar to those seen in transplantation studies using syngeneic mice (Rao et al. 1997; Zanjani et al. 1997; Blomberg et al. 1998). Preliminary data in non-human primates have suggested a similar effect of cell dose (Shields and Andrews 1999). There are two potential concerns related to giving the fetus very high cell numbers as part of in utero transplants. First, the risk of graft-versus-host disease (GvHD) increases as the number of immunocompetent T cells increases. While CD34 selection will significantly deplete donor cells of T cells, further T-cell depletion may be required to bring the absolute number of donor T cells to a level that will prevent GvHD. The one caveat of T-cell depletion is that the rate of graft failure will increase. The optimal percentage and absolute number of donor T cells that can safely be given to the fetus is uncertain at this time. The second potential concern of giving large numbers of donor cells is the recent report of fetal "over-engraftment" (Bambach et al. 1997). It should be noted that in this particular case optimal tissue sampling to rule out GvHD was precluded due to the macerated condition of the fetus and that a total of $2.7 \times 10 \times 8$/kg CD3$^+$ T cells was given to the fetus.

In summary, increasing both the number of donor cells given to the fetus and giving the cells in multiple doses appear to enhance the level of chimerism. Additionally, in the fetal environment, which appears highly competitive, very large numbers of CD34$^+$ cells may be required to achieve clinically relevant levels of chimerism in the absence of an inherent cellular defect (i.e., immunodeficiency).

13.3.3 Sources of Donor Stem Cells

Transplantable HSC are present in fetal liver cells, umbilical cord blood, and adult BM. The number of recoverable HSC from BM and peripheral blood can be enhanced by preharvest treatment of the donor with hematopoietic growth factors. Homogenated fetal liver contains a high frequency of HSC, and the fetal liver cells might exhibit some potential advantages in fetal IUT. They are primed for competing in the fetal environment and, if obtained early enough in gestation, contain no post-thymic T cells (Zanjani et al. 1997). Moreover, they have intense erythroid commitment in the first trimester and would therefore provide an interesting transplant in cases of hemoglobinopathies. In addition, fetal liver cells have a higher turnover rate (Lansdorp 1995), with greater telomere length and proliferation potential than cord blood or BM. Although, it seems possible to organize and retrieve sterile fetal liver cells from 1st trimester abortions (Ek et al. 1994; Westgren et al. 1994), the clinical practice may be hampered due limited availability and ethical considerations in some societies. Most importantly, if postnatal "booster" therapy is going to be used, then HSC must be obtained from a renewable source (Flake and Zanjani 1997; Zanjani et al. 1997). Umbilical cord blood also has the same advantages and limitations as fetal liver cells and, as such, it is unlikely at the present time that they will be useful for clinical trials of in utero stem cell transplantation. Despite these limitations, fetal sources of stem cells should not be discounted completely. As techniques for ex vivo expansion of stem cells improve, the use of fetal sources of stem cells may be preferable, especially in the light of the favorable results from the use of cord blood stem cells in postnatal transplantation.

CD34$^+$ and CD34$^+$/T-cell depleted adult BM offer a number of advantages for IUT. They are a renewable cell source, easily obtained, if necessary HLA matching can be done. CD34$^+$ cells from mobilized peripheral blood or obtained from BM after pretreatment with recombinant hematopoietic growth factors (rhGFs) have not been studied in animal models or in human cases of in utero HSC transplantation. In non-human primates and humans, hematopoietic engraftment is more rapid after postnatal BM transplantation with "mobilized" marrow or peripheral blood HSC. These cells are more proliferative than those from unperturbed BM (Andrews et al. 1992) and they have in vitro proliferative characteristics similar to early gestation fetal progenitors (Thilaganathan et al. 1994; Shields and Andrews 1998).

With the exception of the sheep model, there is relatively little information available suggesting an optimal cell type to be used. Questions related to the use of fetal or adult sources remain to be tested. In addition, questions related to the use of recombinant growth factor-stimulated or -mobilized peripheral blood cells, which have gained popularity in postnatal settings, remain to be tested in non-ovine animal models. Further questions related to the type of cell to enrich remain to be addressed, especially in the light of the recent data suggesting that KDR selection may enrich for developmentally early stem cells (Ziegler et al. 1999).

13.3.4 Technical Aspects of IUT

For delivery of the transplanted cells to the fetus, two principal routes are available: the intraperitoneal and the intravascular. In the sheep model, Zanjani et al. have demonstrated a higher degree of engraftment after intraperitoneal administration (Zanjani 1992). The reason for this is unknown, but it may be that the intraperitoneal cavity can serve as a reservoir for the transplanted cells, with a gradual shedding of cells into the fetal circulation. In the human fetus, Westgren et al. (1997) performed studies with radiolabeled fetal liver cells in vivo in ongoing second trimester abortions. Intravascular administration yielded a significantly higher cell uptake than intraperitoneal in liver, spleen and thymus. With a cell dose of 10^7 cells/kg, we estimated the donor-to-recipient cell ratio of CD34$^+$ cells to be approximately 1/100–1000 with intravascular administration and 1/1000–1/10,000 with intraperitoneal administration. It is an open question whether these cell doses of donor cells are sufficient to provide the donor cells with a competitive edge over the native stem cells.

The technical aspects of the administration of stem cells also include the risk of procedure-related complications (ruptured fetal membranes and infection) and fetal loss rate. IUT is usually carried out using a 22-gauge needle and involves an injection of less than 1 ml. From experience with invasive prenatal diagnosis and therapy including cordocentesis and intraperitoneal administration of blood products, one estimates that the loss rate should be approximately 1%. Intravascular administration probably entails a higher complication rate than intraperitoneal. This is especially true for the very early second trimester

Table 1. Summary of human in utero stem cell transplantations

	Indication	GA (weeks)	Tissue	Cells (n)	Engraft	Comments	Reference
Thalassemia							
1.	α-Thalassemia	13	Paternal BM	3×10⁶/kg CD34+	Yes	Microchimerism noted at birth, evidence of tolerance to donor, transfusion-dependent	Hayward et al. 1998
		19		3×10⁶/kg CD34+			
		24		3×10⁶/kg CD34+; total T cells <1.5×10⁵/kg			
2.	α-Thalassemia	15	FL (n=7; 5–10 weeks)	20.4×10*8/kg	No	Transfusion-dependent	Westgren et al. 1996
		31	FL (n=6; 8–11 weeks)	1.2×10*8/kg			
3.	α-Thalassemia	18	Maternal BM	XX	Yes	Elective termination	Diukman and Golbus 1992
4.	β-Thalassemia	25	Sibling BM	6×10*9/kg	No	Born with disease	Slavin et al. 1992; Touraine et al. 1992
5.	β-Thalassemia	14	FL (9.5 weeks) + thymic cells	1.5×10*10/kg	Yes	0.9% HbA at 6 months	Peschle 1997
6.	β-Thalassemia	14	FL	XX	?	Septic abortion – Detroit	Cowan and Golbus 1994
7.	β-Thalassemia	18	FL (n=5; 6–10 weeks)	8.6×10*8/kg	No	Transfusion-dependent	Westgren et al. 1996
8.	β-Thalassemia	15	FL (n=5; 8–11 weeks)	9×10*8/kg	?	Procedure-related death	Palermo (unpublished)
9.	β-Thalassemia	16	FL (n=6; 9–11 weeks)	8×10*8/kg	?	Elective termination	Palermo (unpublished)
10.	β-Thalassemia	18	FL (not stated)	XX	?	Procedure-related death	Slavin et al. 1992; Touraine et al. 1992
11.	β-Thalassemia	14	BM	2.9×10*8/kg	Yes	At birth, BM=4% metaphases; at 4 months all tests negative	Peschle 1997
		16	HLA matched sibling	2×10*7/kg T cells: 1.7×10*5/kg			
12.	β-Thalassemia	?	Paternal BM	?	No	Transfusion-dependent	Monni et al. 1998

Table 1. Continued

Indication	GA (weeks)	Tissue	Cells (n)	Engraft	Comments	Reference
Red cell disorders						
13. Sickle cell anemia	13	FL (n=5: 6–10 weeks)	16.7×10*8/kg	No	Transfusion-dependent	Westgren et al. 1996
14. Sickle cell anemia	?	BM	?	No	NA	Detroit (unpublished)
15. Sickle cell anemia	?	BM	?	No	NA	Philadelphia (unpubl.)
16. Rh isoimmunization	17	Maternal BM	7.5×10*8/kg	No	Not tolerant to maternal antigens	Linch et al. 1986
17. Rh isoimmunization	12	Maternal BM	1.5×10*10/kg 7.6×10*8/kg=CD34	No	Some tolerance to maternal antigens	Thilagantan and Nicoliades 1993
Immunodeficiencies						
18. BLS	30	FL (7 & 7.5 weeks) and thymic cells	1.2×10*7/kg	Yes	Alive and well	Touraine et al. 1989
19. SCID	28	FL (7.5 weeks)	3.7×10*7/kg	Yes	Alive and well	Touraine et al. 1991
20. SCID (Rak)	23	Paternal BM	1.7×10*7/kg	Yes	At birth negative	Wengler et al. 1986
	24		8.3×10*6/kg		Postbirth retested and positive	Lanfranchi et al. 1998
21 SCID	20	Maternal BM	XX	No	Elective termination	Diukman and Golbus 1992
22. X-SCID	16	Paternal BM	1.1×10*8/kg		Born alive	Flake et al. 1996
	17.5		8.9×10*6/kg		Split chimerism	Flake et al. 1996
	18.5		6.2×10*6/kg			Flake et al. 1996
23. X-SCID	21	Paternal BM	3.9×10*7/kg	Yes	Born alive	Wengler et al. 1996
	22		1.1×10*7/kg T cells: 3.4×10*5/kg		Split chimerism	
24. X-SCID	22	Paternal BM	6×10*6CD34+	Yes	Born alive, split chimerism	Lanfranchi et al. 1998
25. X-SCID	14	FL (10 wk)	7×10*7/kg	Yes	Born alive, split chimerism	Westgren 1999 (unpubl.)
26. CGD	17, 21	FL (not stated)	XX	?	Procedure-related death	Touraine 1996
27 CGD	13.5	BM CD34+	2.9×10*8/kg	No	Born with disease; no chimerism	Harrsion 1998 (unpubl.)
28 Chediak-Higashi	19	Maternal BM	XX	No	Born with disease; no chimerism	Flake and Zanjani 1997
29 Chediack-Higashi	19	Maternal BM	XX	No	Born with disease; no chimerism	Diukman and Golbus 1992

Table 1. Continued

	Indication	GA (weeks)	Tissue	Cells (n)	Engraft	Comments	Reference
	Storage disease						
30	Globoid cell Leukodystrophy	14	Paternal BM	CD34+: 5×10*9/kg CD3+: 2.7×10*8/kg	Yes	Fetal death at 20 weeks	Bambach et al. 1997
31	Globoid cell leukodystrophy	13	Paternal BM	5×10*8/kg	No	Born with disease; low degree of split chimerism	Leung et al. 1999
32	Globoid cell leukodystrophy	13	Paternal BM	5×10*8/kg	No	Born with disease; no chimerism	Leung et al. 1999
33	Hurler's syndrome	14	FL	?	No	Died of disease	Cowan and Golbus 1994
34	Niemann–Pick disease	12, 13	FL (not stated)	XX	?	No data	Touraine 1996
35	Metachromatic leukodystrophy	37	Paternal BM	1×10*9/kg	No	Born with disease; no chimerism	Slavin et al. 1992; Touraine et al. 1992
36	Metachromatic leukodystrophy	23	Paternal BM	4.3×10*9/kg	No	born with disease; no chimerism	Slavin et al. 1992; Touraine et al. 1992

BLS, bare lymphocyte syndrome; *CGD*, chronic granulomatous disease; *SCID*, severe combined immunodeficiency disease; *X-SCID*, X-linked SCID.

fetuses. It is noticeable that 4 of 37 known attempts to perform fetal transplantation have ended in procedure-related deaths (Table 1).

13.3.5 The Role of Gestational Age on In Utero Engraftment

In both ovine and non-human primate models, the gestational age at the time of transplantation correlates with the rate of engraftment. In the ovine model, engraftment has been demonstrated over a large gestational age window (0.25–0.83 gestation). However, engraftment and chimerism were higher when the transplant was performed between 0.35 and 0.45 gestation and then declined at gestational ages of .35 and 0.50 (Zanjani 1992; Zanjani et al. 1997). Roodman et al. (1991), using fetal baboons, noted that no measurable chimerism was achieved when the transplantation occurred after 0.45 gestation. Why there is a drop in the rate of engraftment in the fetal sheep model below 0.35 gestation is unclear, as donor cells have been noted to "home" to both the fetal liver and BM, depending on the gestational age at the time of the transplant in both sheep and primate models (Zanjani et al. 1992; Cowan and Golbus 1994). Whether cells that engraft into the fetal liver are then unable to migrate to the fetal BM remains untested. If this were true, then transplants done at gestational ages prior to available hematopoietic niches within the BM would result in non-engraftment. Failure of engraftment after 0.45 gestation has been assumed to be related to the development of a partially functional fetal immune system and the transplantation of partially or completely HLA mismatched donor cells (Jones et al. 1996; Flake and Zanjani 1997; Zanjani et al. 1997). If the fetal immune system is the limiting factor with regard to successful engraftment, then the window of immune opportunity in the human and non-human primate fetus is likely to be earlier (6–12 weeks) than previously suspected (12–16 weeks; Van Furth et al. 1965; Auguset 1971; Stites et al. 1974; Gale 1987; Jaleco et al. 1997; Stark et al. 1991; Lindton et al., in press).

13.3.6 The Role of HLA on the Rate of In Utero Engraftment

In postnatal BM transplantation, HLA matching is essential for successful engraftment without GvHD. The rate of graft failure and GvHD is also directly related to the degree of HLA mismatch (Jankowski and Ildstadt 1997). The few experiments of nature (marmosets and cows) and the ovine model of IUT have demonstrated that successful engraftment with clinically relevant levels of chimerism can occur even when significant HLA barriers appear to exist. Based on this data, it is presumed that successful fetal engraftment of HSC will occur with less than perfect HLA matching due to the incompetent fetal immune system. Additionally, if the in utero transplant is done early enough in gestation, the donor cells should induce immune tolerance and be recognized as "self" (Touraine 1990; Crombleholme et al. 1991; Diukman and Golbus 1992; Hajdu and Golbus 1993; Zanjani et al. 1997). Unfortunately, the levels of chimerism noted in marmosets, cows, and sheep have not been repeated in published reports of non-human primates or in human trials of in utero stem cell transplantation in the absence of immune deficiencies. To date, the role that HLA plays in the success of in utero HSC transplantation has not been addressed. Based on studies of postnatal stem cell transplantation, it is reasonable to assume that even in the setting of IUT, HLA may have a significant role in determining the rate of engraftment and the level of chimerism. Examination of the role of HLA in fetal transplantation deserves further investigation.

13.3.7 In Utero Induction of Tolerance and Postnatal Boosting to Enhance Chimerism

Experiments of nature and animal models suggest that tolerance induction occurs with successful in utero engraftment of donor HSC, even if the level of chimerism is very low (Carrier et al. 1995; Cowan et al. 1996). Genetic diseases affecting the lympho-hematopoietic system can be categorized into those that result in significant damage to the developing fetus prior to birth (storage diseases, leukodystrophies, and α-thalassemia) and those that do not affect the fetus but result in significant morbidity postnatally (β-thalassemia, sickle cell disease, and immune deficiencies). For those diseases that affect the fetus prior to birth, clinically effective levels of chimerism must be obtained in utero. For diseases that express their deleterious effects postnatally, induction of tolerance may allow more efficacious postnatal therapy, without the usual risk associated with pretransplantation conditioning (chemotherapy, immune suppression, and radiation therapy). Although, data from animal models are limited, they are also encouraging. Induction of tolerance (skin grafting and solid organ transplantation) and increased hematopoietic chimerism with postnatal booster dosing have been demonstrated in a limited number of mice (Carrier et al. 1995), sheep (Zanjani et al. 1994), and non-human primates (Cowan et al. 1996). Two cases of human intrauterine transplantation have been reported where donor-specific tolerance was tested. In both of these cases, a reduced response to the donor was noted in MLC and in cytotoxic T-lymphocyte testing (Thilaganthan and Nicoliades 1993; Hayward et al. 1998). If further study of the response to postnatal boosting demonstrates that clinically relevant levels of chimerism can be achieved, then the number of candidate diseases for IUT could be greatly expanded.

13.4 In Utero Transplantation in Man

Thirty-seven cases of allogeneic in utero stem cell transplantation in humans have been either reported or presented at national and international meetings. The majority of these cases were recently reviewed (Jones et al. 1996; Flake and Zanjani 1999) and are summarized in Table 1. In many of these cases, details relating to the individual cases are limited. The source of donor cells has been HLA-mismatched fetal liver cells, T-cell depleted partially HLA-matched adult BM (with the majority of donors being one of the parents, and 1 case of a HLA-matched sibling), and HLA haploidentical CD34$^+$ enriched cells obtained from one of the parents. Detectable engraftment has been reported in 13 of the 37 cases. In all cases, with the exception of the severe immunodeficiencies, the level of chimerism has been very low and had no impact on the course of disease being treated. In the 4 well-documented cases of in utero therapy of X-SCID, one could argue that therapy postnatally, without conditioning, would have been equally effective (Buckley et al. 1999). However, it should be noted that these infants were born with functional T cells, reducing the risk of postnatal opportunistic infection. In 1 of these cases, tested for B-cell function with a unique antigen (bacteriophage φ 174), the infant showed near normal immunoglobulin (IgM, IgG) response (Dr. Ochs, personal communication). Infants treated postnatally have not shown this type of response with pretransplant conditioning, suggesting that in utero therapy may offer some advantages over immediate postnatal therapy. Additional cases of in utero-treated X-SCID will be needed to fully evaluate this effect.

It is unclear whether the failure of most in utero allogeneic HSC transplants in humans is related to fetal immune function, HLA mismatching of donor cells, a lack of potential space or niches for the donor cells to engraft, the types and numbers of stem cells transplanted, or the inability of donor cells to traffic to sites of engraftment. The lack of information explaining the reason for the failed cases of human in utero stem cell transplantation leaves unresolved questions. These data further suggest that additional attempts at in utero stem cell transplantation, for non-immunological diseases, should not be carried out until additional animal data suggest that alternative therapeutic approaches are likely to improve on the dismal results published to date.

13.5 New Directions

The failure of in utero stem cell transplantation, except in the cases of severe immunodeficiencies, suggest that the current techniques are not appropriate for this type of therapy and that future research should be directed at techniques that will potentially enhance the success of fetal stem cell therapy. Some of these areas are addressed in the next section.

13.5.1 Myeloablation and Immunosuppression

Successful postnatal BM transplantation requires the patient to undergo pretransplantation chemotherapy and/or irradiation for ablation as well as pre- and post-transplantation immune suppression. As noted above, the only clinically successful in utero transplants have occurred in the setting of severe immune deficiencies. This data would suggest that the fetal immune system plays a vital role in the success of in utero transplants. Alternatively, the success of transplants in fetuses with severe immune deficiencies may be related to increased hematopoietic space that is available when certain lineages of fetal cells (T cells and NK cells in the case of X-SCID) are absent.

An important argument for fetal transplantation has been that engraftment can occur in normal fetuses without the use of cytoablative
drugs or immunomodulation. However, in recent studies this concept
has been questioned, since engraftment in fetuses with normal immunological function have failed, and experimental studies have revealed a
response to allogeneic stimulation as early as the 12th week of gestation
(Lindton et al., in press). Thus, ablation and immunosuppression might
be considered. The rationale to do this would be twofold: to create space
for homing of donor dells and to inhibit an allogeneic response.

Different chemotherapeutic agents could be considered such as Al-
ceran, ARA-C, Daunorubricin, Doxorubricin, etc. All of these drugs
have been used in pregnant women with malignancies with favorable
fetal outcome (Zemlickis et al. 1992; Randall 1993). However, most
studies on fetal outcome after in utero exposure have revealed an increased likelihood of spontaneous abortions and malformations when
chemotherapy is used in the first trimester, whereas the risk is not
apparent beyond the first trimester. Available data are mostly based on
case reports and literature reviews and could, for obvious reasons, be
biased. Available information does not permit any quantitative conclusions regarding effects on the fetus, and one can really question whether
one will ever purposefully expose the human fetus to such drugs. At the
Karolinska Institutet in Stockholm, we have during recent years performed in vitro studies on fetal hematopoietic colony formation and
MLC after exposure to the above-mentioned chemotherapeutic agents
and some other drugs with known immunomodulative effects such as
corticosteroids, anti-CD3 monoclonal antibody (OKT-3) and antithy-
mocyte antigen (ATG). Of these drugs, betamethasone was the most
promising alternative in respect to the inhibitory effect of fetal stem
cells, inhibition of MLC, and applicability.

Due to the potential harmful effects of conventional chemotherapy,
we have explored alternative new methods for the inhibition of fetal
hematopoiesis, and thereby found a selective advantage of donor cells at
the time of in utero transplantation. Recently we could demonstrate that
the B19 parvovirus capsid exhibits in vitro a 70–95% reduction of burst
forming unit-erythroid cell, colony forming units-granulocytes,
erythroid cells, macrophages and megakaryocytes. This might be a
strategy to down-regulate indigenous hematopoiesis in recipients prior
to stem cell transplantation, and we are therefore now embarking on in

vivo studies in a primate model. Another strategy that is explored by researchers in this field is monoclonal antibodies against surface markers of fetal HSC such as CD34 class ll (Jones, personal communication).

To summarize, different strategies for down-regulation of indigenous hematopoiesis and immunomodulation are presently being explored. These strategies for enhancement of engraftment have not only implications for fetal stem cell transplantation, but also for other alternative strategies such as gene therapy.

13.5.2 Fetal Gene Therapy

Gene transfer into autologous HSC may be an alternate therapeutic approach to allogeneic HSC transplantation if stable gene integration and expression can be obtained in a large enough proportion of stem cells and their progeny. In large animals and in humans transplanted postnatally, in vivo gene transfer efficiency has been very low. Fetal HSC appear to be more susceptible to retroviral mediated gene transfer in vitro compared with adult progenitor cells (Ekhterae et al. 1990). Gene transfer into fetal HSC could be approached in two ways. First, autologous fetal HSC could be harvested, transduced in vitro, and then transplanted in utero into the expanding fetal hematopoietic environment. This is based on the assumption that sufficient numbers of fetal HSC can be collected. Second, fetal HSC might be transduced in vivo by introducing vector or vector-producing cells into the fetus with subsequent in vivo transduction of hematopoietic tissues (Shields et al. 2000) – the caveat being that there will be a risk of transduction of non-hematopoietic cells, including germ cells. The earlier in gestation that such therapy can be achieved, the greater the potential for developing tolerance to potentially immunogenic novel proteins expressed by modified cells (Silverstein et al. 1963; Peeters et al. 1996; Riddell et al. 1996). Additionally, gene therapy of fetal HSC may be beneficial even if the desired gene product is produced transiently. This may prevent or reduce in utero damage that occurs in many storage diseases, thereby allowing postnatal therapies to be more effective. At the present time it remains to be determined whether sufficient fetal HSC can be collected for autologous transplantation and whether these cells will be suitable targets for gene transfer. In addition, at the present time there is insufficient evidence to suggest either the safety or efficacy of in utero gene therapy, and future efforts should be directed at preclinical models in both small and large animals.

13.6 Ethical Considerations in the Context of What Diseases Should Be Targeted by Intrauterine Fetal Transplantations

The fetus will become a patient by the informed agreement and consent of the pregnant women. Other attempts to argue for an independent moral status of the fetus or that the fetus has no moral status at all fail on rational grounds. For practical purposes, it appears that the prevailing practice in fetal medicine is to regard the pregnant woman as the moral fiduciary of the patient. Introduction of a new modality for fetal therapy is afflicted with several inherent difficulties (Fletcher 1992; McCullough and Chervnaek 1994). Major ethical difficulties are risk/benefit assessments, selection of cases for treatment, and optimizing informed consent. The most frequent and difficult ethical problem is the selection of the patient, because of the lack of knowledge about what outcome can be expected and which case will benefit from the procedure. With regard to informed consent, precautions are clearly needed due to the vulnerability of the pregnant women and the enthusiasm for treatment the responsible physician may express. All these ethical considerations affect in utero fetal transplantations.

In the present review, we have tried to summarize all cases of human fetal stem cell transplantations. The transplantations have been carried out in a wide range of disorders ranging from treatable conditions during fetal life such as rhesus disease to non-curable lethal disorders such as α-thalassemia. Considering attempting to perform a new innovative treatment such as IUT with unproven expectations involves an assessment of the pathophysiology of the condition to be treated and up-to-date knowledge about previous experience with pre- and postnatal stem cell transplantations. It is of the utmost importance that cases selected for these experimental treatments should have an acceptable risk/benefit ratio. Furthermore, since experience will be scarce in all centers, it is a prerequisite that all cases are reported. Attempts to establish Internet-based registers are in progress.

Table 2. Candidate diseases

Hematopoietic disorders	
Disorders affecting lymphocytes	
SCID (sex-linked)	Agammaglobinemia
SCID (adenosine deaminase deficiency)	Bare lymphocyte syndrome
Disorders affecting erythrocytes	*Disorders affecting granulocytes*
Sickle cell disease	Chronic granulomatous disease
α-Thalassemia	Infantile agranulocytosis
β-Thalassemia	Neutrophil membrane GP-180
Hereditary spherocytosis	Lazy leukocyte syndrome
Fanconi anemia	Chediak-Higashi syndrome
Inborn errors of metabolism	
Mucopolysaccharidoses	*Mucolipidoses*
MPS I (Hurler Disease)	Gaucher disease
MPS II (Hunter Disease)	Metachromatic leukodystrophy
MPS IIIB (Sanfilippo B)	Krabbe disease
MPS IV (Morquio)	Niemann-Pick disease
MPS VI (Maroteaux-Lamy)	β-Glucoronidase deficiency
	Fabry disease
	Adrenal leukodystrophy
Mannosidosis	
α-Mannosidosis	
β-Mannosidosis	

Proposed candidates for in utero treatment are listed in Table 2. Treatment with in utero HSC in cases of immunodeficiency syndrome has yielded encouraging results both using T-cell depleted BM and fetal liver cells. Successful in utero therapy of X-SCID has been reported from different centers. It seems that it is the fundamental defect in SCID cases with defective cell signaling that will not facilitate proliferation of indigenous cells that makes them exquisitely amenable to in utero therapy. It is an open question whether IUT has advantages over conventional postnatal BM transplantations. Potential advantages with IUT include reconstitution of T cells prenatally, no postnatal period of susceptibility for infections, simplicity of prenatal treatment and the relative cost-saving. Whether B-cell function is improved over what can be achieved by postnatal therapy remains to documented. Other forms of SCID that might benefit from IUT are adenosine deficiency, Bruton disease, and chronic granulomatous disease. All these conditions differ with respect to pathophysiology, and experience needs to be gathered for each category of patients. However, even though there is a significant immunological defect in each of these conditions, complete absence of lineages (T cells, B cells, granulocytes) or defects are not present, which may limit successful in utero therapy using currently reported techniques. Until additional animal studies suggest strategies that may prove the effectiveness of in utero therapy in conditions with less severe immune defects, human attempts at in utero correction do not appear to be justified.

Considering the current results in cases with hemoglobinopathies and inborn errors of metabolism where in utero transplantation have convincingly failed, we question the value of continuing to perform conventional in utero stem cell transplantation. Other strategies as outlined above might be attempted. One could argue that conditions where the outcome will be unequivocally poor, with fetal death or where no cure exists after birth such as in α-thalassemia, Sly syndrome and Wolman disease, could be candidates, since everything is to be gained and little to be lost.

Finally, the complexity of fetal stem cell transplantation calls for collaboration between many fields in medicine. A multidisciplinary approach is clearly a prerequisite if progress is going to be seen in this field. It is necessary that this new potentially valuable treatment be performed and assessed in centers with adequate facilities for both basic and clinical science, and where adequate follow-up of engraftment and clinical outcome can be conducted.

Acknowledgements. Supported by the Swedish Medical Research Council (K99–72X-11231–05 C) and the Children's Cancer foundation (1995/035).

References

Andrews RG, Bensinger WI, Knitter GH et al (1992) The ligand for c-kit, stem cell factor, stimulates the circulation of cells that engraft lethally irradiated baboons. Blood 80:2715–2720

Auguset C (1971) Onset of lymphocyte function in the developing human fetus. Pediatr Res 5:539–547

Bambach BJ, Moser HW, Blakemore K et al (1997) Engraftment following in utero bone marrow transplantation for globoid cell leukodystrophy. Bone Marrow Transplant 19:399–402

Blomberg M, Rao S, Reilly J et al (1998) Repetitive bone marrow transplantation in nonmyeloablated recipients. Exp Hematol 26:320–324

Buckley RH, Schiff SE, Schiff RI et al (1999) Hematopoietic stem cell transplantation for the treatment of severe combined immunodeficiency. N Engl J Med 340:508–516

Carrier E, Lee TH, Busch MP, Cowan MJ (1995) Induction of tolerance in nondefective mice after in utero transplantation of major histocompatibility complex-mismatched fetal hematopoietic stem cells. Blood 86:4681–4690

Cowan MJ, Golbus M (1994) In utero hematopoietic stem cell transplants for inherited diseases. Am J Pediatr Hematol Oncol 16:35–42

Cowan MJ, Tarantal AF, Capper J, Harrison M, Garovoy M (1996) Long term engraftment following in utero T cell depleted parental marrow transplantation into fetal rhesus monkeys. Bone Marrow Transplant 17:1157–1165

Crombleholme TM, Harrison MR, Zanjani ED (1990) In utero transplantation of hematopoietic stem cells in sheep: the role of T cells in engraftment and graft-versus-host disease. J Pediatr Surg 25:885–892

Crombleholme TM, Langer JC, Harrison MR, Zanjani ED (1991) Transplantation of fetal cells. Am J Obstet Gynecol 164:218–229

Diukman R, Golbus M (1992) In utero stem cell therapy. J Reprod Med 37:515–520

Ebert U, Loffler H, Kirch W (1997) Cytotoxic therapy and pregnancy. Pharm Ther 74:207–220

Ek S, Westgren M, Ringden O et al (1994) Infectious screening in fetal stem cell collection. Fetal Diagn Ther 9:357–362

Ekhterae D, Crumbleholme T, Karson E, Harrison MR, Anderson WF, Zanjani ED (1990) Retroviral vector-mediated transfer of the bacterial neomycin resistance gene into fetal and adult sheep and human hematopoietic progenitors in vitro. Blood 75:365–369

Flake A (1986) Transplantation of fetal hematopoietic stem cells in utero: the creation of hematopoietic chimeras. Science 223:776–778

Flake AW, Zanjani ED (1997) In utero hematopoietic stem cell transplantation. A status report. JAMA 278:932–937

Flake AW, Zanjani E (1999) In utero hematopoietic stem cell transplantation: ontogenic opportunities and biological barriers. Blood 54:2179–2191

Flake A, Harrison M, Zanjani E (1991) In utero stem cell transplantation. Exp Hematol 19:1061–1064

Flake AW, Roncarolo M-G, Puck JM (1996) Treatment of X-linked severe combined immunodeficiency in utero transplantation of paternal bone marrow. N Engl J Med 335:1806–1810

Fletcher JC (1992) Fetal therapy, ethics and public policy. Fetal Diagn Ther 7:158–168

Gale R (1987) Development of the immune system in human fetal liver. Thymus 10:45–56

Gathings WE, Lawton AR, Cooper MD (1977) Immunofluorescent studies of the development of pre-B cells,B lymphocytes and immunoglobulin isotype diversity in humans. Eur J Immunol 7:804–810

Hajdu K, Golbus MS (1993) Stem cell transplantation. West J Med 159:356–359

Harrison M (1989) In-utero transplantation of fetal liver haemopoietic stem cells in monkeys. Lancet 2:1425–1427

Hayward A, Ambruso D, Battaglia F et al (1998) Microchimerism and tolerance following intrauterine transplantation and transfusion for alpha-thalassemia-1. Fetal Diagn Ther 13:8–14

Jaleco AC, Blom B, Res P et al (1997) Fetal liver contains committed NK progenitors, but is not a site for development of CD34+ cells into T cells. J Immunol 159:694–702

Jankowski RA, Ildstad ST (1997) Chimerism and tolerance: from freemartin cattle and neonatal mice to humans. Hum Immunol 52:155–161

Jones DR, Bui TH, Anderson EM et al (1996) In utero haematopoietic stem cell transplantation: current perspectives and future potential. Bone Marrow Transplant 18:831–837

Lanfranchi A, Neva A, Tettoni K, Veradi R, Mazzolari E, Wengler G et al (1998) In utero transplantation (IUT) of parental CD34+ cells in patients affected by primary immunodeficiencies. Bone Marrow Transplant 21:S127

Lanier LL (1995) The role of natural killer cells in transplantation. Curr Opin Immunol 7:626–631

Lansdorp PM (1995) Telomere length and proliferation potential of hematopoietic stem cells. Stem Cell Trans 108:1–6

Leung W, Blakemore K, Jones R-J et al (1999) A human murine chimera for in utero human hematopoietic stem cell transplantation. Biol Blood Marrow Transpl 5:1–7

Linch D, Rodeck C, Nicolaides K, Jones H, Brent L (1986) Attempted bone-marrow transplantation in a 17-week fetus. Lancet 27:1453

Lindton B, Markling L, Ringden O, Kjaeldgaard A, Gustafsson O, Westgren M (2000) Mixed lymphocyte culture of human fetal liver cells. Fetal Diag Ther 15:71–78

McCullough L, Chervenaek F (1994) Ethics in obstetrics and gynecology. Oxford University Press, New York

Monni G, Ibba RM, Zoppi MA, Floris M (1998) In utero stem cell transplantation. Croat Med 39:220–223

Muench MO, Cupp J, Polakoff J, Roncarolo MG (1994) Expression of CD33, CD38, and HLA-DR on CD34+ human fetal liver progenitors with a high proliferative potential. Blood 83:3170–3181

Opie TM, Shields LE, Andrews RG (1998) Cell surface antigen expression in early and term gestation fetal hematopoietic progenitor cells. Stem Cells 16:343–348

Peeters MJ, Patijn GA, Lieber A, Meuse L, Kay MA (1996) Adenovirus-mediated hepatic gene transfer in mice: comparison of intravascular and biliary administration. Hum Gene Ther 7:1693–1699

Peschle C (1997) In utero transplantation of purified stem cells from an HLA identical sibling into a b-thalassemia embryo. In utero stem cell transplantation and gene therapy, 2nd international meeting, Nottingham

Randall T (1993) National registry seeks scarce data on pregnancy outcomes during chemotherapy. JAMA 260:323

Rao SS, Peters SO, Crittenden RB et al (1997) Stem cell transplantation in the normal nonmyeloablated host: relationship between cell dose, schedule, and engraftment. Exp Hematol 25:113–121

Riddell SR, Elliott M, Lewinsohn DA et al (1996) T-cell mediated rejection of gene-modified HIV-specific cytotoxic T lymphocytes in HIV-infected patients (see comments). Nat Med 2:216–223

Roodman GH, Kuehl TJ, Vanderberg JL, Muirhead DY (1991) In utero bone marrow transplantation of fetal baboons with mismatched adult bone marrow. Blood Cells 17:367–375

Shields L, Andrews R (1999) In utero engraftment by allogenic CD34+ marrow cells in nonhuman primates: relationship to the absolute number of cells transplanted. J Soc Gynecol Invest 6:186A

Shields LE, Andrews RG (1998) Gestational age changes in circulating CD34+ hematopoietic stem/progenitor cells in fetal cord blood. Am J Obstet Gynecol 178:931–937

Shields LE, Kiem H-P, Andrews RG (2000) A comparison of preterm and term gestation umbilical cord blood CD34+ hematopoietic progenitor cells (HPC) as targets for retroviral mediated gene transfer. Am J Obstet Gynecol (in press)

Silverstein A, Uhr J, Kraner K (1963) Fetal Response to antigenic stimulus: antibody production by the fetal lamb. J Exp Med 117:799–812

Slavin S, Naparstek E, Ziegler M, Lewin A (1992) Clinical application of intrauterine bone marrow transplantation for treatment of genetic diseases – feasibility studies. Bone Marrow Transplant 9:189–190

Stark JH, Smit JA, Neethling FA, Nortman PJ, Myburgh JA (1991) Immunological compatibility between the chacma baboon and man. Transplantation 52:1072–1078

Stites DP, Carr MC, Fudenberg HH (1974) Ontogeny of cellular immunity in the human fetus: development of responses to phytohemagglutinin and to allogenic cells. Cell Immunol11:257–271

Surbek DV, Gratwohl A, Holzgreve W (1999) In utero hematopoietic stem cell transfer: current status and future strategies. Eur J Obstet Gynecol 85:109–115

Tavassoli M (1991) Embryonic and fetal hemopoiesis: an overview. Blood Cells 1:269–281

Thilaganthan B, Nicoliades K (1993) Intrauterine bone-marrow transplantation at 12 gestation. Lancet 342:243

Thilaganathan B, Abbas A, Nicolaides KH (1993) Fetal blood natural killer cells in human pregnancy. Fetal Diagn Ther 8:149–153

Thilaganathan B, Nicolaides KH, Morgan G (1994) Subpopulations of CD34-positive haemopoietic progenitors in fetal blood. Br J Haematol 87:634–636

Touraine JL (1990) In utero transplantation of stem cells in humans. Nouv Rev Fr Hematol 32:441–444

Touraine JL (1996) Treatment of human fetuses and induction of immunological tolerance in humans by in utero transplantation of stem cells into fetal recipients. Acta Haematol 96:115–119

Touraine JL, Raudrant D, Royo C et al (1989) In utero transplantation of stem cells in the bare lymphocyte syndrome. Lancet i:1382

Touraine JL, Raudrant D, Royo C et al (1991) In utero transplantation of fetal liver stem cells in humans. Blood Cells 17:379–387

Touraine JL, Raudrant D, Rebaud A et al (1992) In utero transplantation of stem cells in humans: immunological aspects and clinical follow-up of patients. Bone Marrow Transplant 1:121–126

Van Furth R, Schuit H, Hijmans W (1965) The immunological development of the human fetus. J Exp Med 122:1173–1187

Wengler GS, Lanfranchi A, Frusca T et al (1996) In-utero transplantation of parental CD34 haematopoietic progenitor cells in a patient with X-linked severe combined immunodeficiency (SCIDXI). Lancet 348:1484–1487

Westgren M, Ek S, Bui TH et al (1994) Establishment of a tissue bank for fetal stem cell transplantation. Acta Obstet Gynecol Scand 73:385–390

Westgren M, Ek S, Jansson B et al (1997) Tissue distribution of transplanted fetal liver cells in the human fetal recipient. Am J Obstet Gynecol 176:49–53

Westgren M, Ringden O, Eik Nes S et al (1996) Lack of evidence of permanent engraftment after in utero fetal stem cell transplantation in congenital hemoglobinopathies. Transplantation 61:1176–1179

Westgren M, Ringden OI (1994) Fetal to fetal transplantation. Acta Obstet Gynecol Scand 73:371–373

Zanjani E (1992) The fetus as an optimal donor and recipient of hemopoietic stem cells. Bone Marrow Transplant 10:107–114

Zanjani ED, Mackintosh FR, Harrison MR (1991) Hematopoietic chimerism in sheep and nonhuman primates by in utero transplantation of fetal hematopoietic stem cells. Blood Cells 17:349–363

Zanjani ED, Ascensao JL, Tavassoli M (1992) Homing of liver-derived hemopoietic stem cells to fetal bone marrow. Trans Assoc Am Phys 105:7–14

Zanjani ED, Ruthven A, Ruthven J et al (1994) In utero hematopoietic stem cell transplantation results in donor specific tolerance and facilitates postnatal boosting of donor cell levels. Blood 84:100–12

Zanjani ED, Almeida-Porada G, Ascensano JL et al (1997) Transplantation of hematopoietic stem cells in utero. Stem Cells 15:79–93

Zemlickis D, Lishner M, Degendorfer P et al (1992) Fetal outcome after in utero exposure to cancer chemotherapy. Arch Intern Med 152:573–576

Ziegler BL, Valtieri M, Porada GA et al (1999) KDR receptor: a key marker defining hematopoietic stem cells. Science 285:1553–1558

Subject Index

Ernst Schering Research Foundation Workshop

Editors: Günter Stock
Monika Lessl

Vol. 1 *(1991)*: Bioscience ⇋ Society – Workshop Report
Editors: D. J. Roy, B. E. Wynne, R. W. Old

Vol. 2 *(1991)*: Round Table Discussion on Bioscience ⇋ Society
Editor: J. J. Cherfas

Vol. 3 *(1991)*: Excitatory Amino Acids and Second Messenger Systems
Editors: V. I. Teichberg, L. Turski

Vol. 4 *(1992)*: Spermatogenesis – Fertilization – Contraception
Editors: E. Nieschlag, U.-F. Habenicht

Vol. 5 *(1992)*: Sex Steroids and the Cardiovascular System
Editors: P. Ramwell, G. Rubanyi, E. Schillinger

Vol. 6 *(1993)*: Transgenic Animals as Model Systems for Human Diseases
Editors: E. F. Wagner, F. Theuring

Vol. 7 *(1993)*: Basic Mechanisms Controlling Term and Preterm Birth
Editors: K. Chwalisz, R. E. Garfield

Vol. 8 *(1994)*: Health Care 2010
Editors: C. Bezold, K. Knabner

Vol. 9 *(1994)*: Sex Steroids and Bone
Editors: R. Ziegler, J. Pfeilschifter, M. Bräutigam

Vol. 10 *(1994):* Nongenotoxic Carcinogenesis
Editors: A. Cockburn, L. Smith

Vol. 11 *(1994)*: Cell Culture in Pharmaceutical Research
Editors: N. E. Fusenig, H. Graf

Vol. 12 *(1994):* Interactions Between Adjuvants, Agrochemical
and Target Organisms
Editors: P. J. Holloway, R. T. Rees, D. Stock

Vol. 13 *(1994):* Assessment of the Use of Single Cytochrome
P450 Enzymes in Drug Research
Editors: M. R. Waterman, M. Hildebrand

Vol. 14 *(1995):* Apoptosis in Hormone-Dependent Cancers
Editors: M. Tenniswood, H. Michna

Vol. 15 *(1995):* Computer Aided Drug Design in Industrial Research
Editors: E. C. Herrmann, R. Franke

Vol. 33 (2001): Stem Cells from Cord Blood, In Utero Stem Cell Development, and Transplantation-Inclusive Gene Therapy
Editors: W. Holzgreve, M. Lessl

Supplement 1 (1994): Molecular and Cellular Endocrinology of the Testis
Editors: G. Verhoeven, U.-F. Habenicht

Supplement 2 (1997): Signal Transduction in Testicular Cells
Editors: V. Hansson, F. O. Levy, K. Taskén

Supplement 3 (1998): Testicular Function:
From Gene Expression to Genetic Manipulation
Editors: M. Stefanini, C. Boitani, M. Galdieri, R. Geremia, F. Palombi

Supplement 4 (2000): Hormone Replacement Therapy
and Osteoporosis
Editors: J. Kato, H. Minaguchi, Y. Nishino

Supplement 5 (1999): Interferon:
The Dawn of Recombinant Protein Drugs
Editors: J. Lindenmann, W.D. Schleuning

Supplement 6 (2000): Testis, Epididymis and Technologies
in the Year 2000
Editors: B. Jégou, C. Pineau, J. Saez